ASA STUDIES

General Editor: Edwin Ardener

1

Biosocial Anthropology

ASA STUDIES

First titles

Biosocial Anthropology

Edited by
Robin Fox

A HALSTED PRESS BOOK

John Wiley & Sons
New York

First published in 1975
by Malaby Press Limited
Aldine House, 26 Albemarle Street, London W1X 4QY

This book has been set in 11 on 12 point Times New Roman
PRINTED IN GREAT BRITAIN

Published in the U.S.A.
by Halsted Press, a Division
of John Wiley & Sons, Inc.,
New York

Library of Congress Cataloging in Publication Data
Main entry under title:

Biosocial anthropology.

(ASA studies; 1)
"A Halsted Press book."
Papers mainly derived from material presented at
a conference sponsored by the Association of Social
Anthropologists of the Commonwealth which was held
at St. John's College, Oxford, 4–11 July 1973.
Bibliography: p.
Includes index.
1. Ethnology—Congresses. 2. Genetic psychology—
Congresses. 3. Social interaction—Congresses.
I. Fox, Robin, 1934–   II. Association of Social
Anthropologists of the Commonwealth. III. Series:
Association of Social Anthropologists of the Common-
wealth. ASA studies; 2.
GN320.B55 1975   573   75-4110
ISBN 0-470-27033-0

To Sherwood L. Washburn
and Nikolaas Tinbergen
this book is respectfully
and affectionately dedicated
by the contributors

Contents

General Editor's Note

ASA Studies is a new series largely based upon selected sessions of the special Decennial Conference of the Association of Social Anthropologists of the Commonwealth, entitled, 'New Directions in Social Anthropology', and held at St John's College, Oxford, 4–11 July 1973.

The last special conference had taken place at Jesus College, Cambridge, in 1963, and a preliminary word should be said about it, to provide a suitable perspective. Those who convened it had, as they later wrote, two main purposes in mind: to provide a 'joint, organized stock-taking' between British and American social anthropologists, and to reflect 'the problems and views of a younger generation of anthropologists'. It was pointed out that between 1946 and 1962 the Association of Social Anthropologists had increased its membership from under a score to over one hundred and fifty. In retrospect, the 1963 Cambridge Conference can be seen to represent on the one hand a kind of coming-of-age of the ASA as an organization (in its rather awkward seventeenth year) and, on the other, an awareness of changes to come in the aims, interests, and personnel of the formerly close-knit British branch of the subject. The four volumes that emerged from that Conference were entitled *The Relevance of Models for Social Anthropology*, *Political Systems and the Distribution of Power*, *Anthropological Approaches to the Study of Religion*, and *The Social Anthropology of Complex Societies* (published by Tavistock, 1964–6) each with a general Introduction by Max Gluckman and Fred Eggan, from which my quotations here and later come.

It was that occasion which provided the base-line for the 1973 Decennial Conference. The ASA committee, under the Chairmanship of Professor Meyer Fortes, wished the 1973 Conference to play the same kind of role in its period as the previous one had done in its own time. The task was, however, quite complicated. The membership of one hundred and fifty or so of ten years before had now risen to several hundreds, while the Conference was this time to be international, not merely Anglo-American. Furthermore, although 'new directions' were to be emphasized as far as possible, subjects which had in recent years been covered by the regular annual ASA conferences (and thus published) or were already scheduled for the future, were to be excluded. Eventually the Committee chose eleven topics, each to be covered in a half-day session, and invited a convener to organize each of them. Finally, the Conference was placed by the Committee under my general convenership. This, I should emphasize, was seen as largely an organizational task, which I have been invited to continue here as General Editor of these volumes. The session-conveners had a free hand in the choice of contributors on their topics, within certain limits of numbers and finance, and each session-convener decided the structure of his own half-day session.

An immediate difference between this conference and the 1963 one lies in this much looser 'federal' structure. The Committee were of the opinion that all sessions should be consecutive, so that all could be attended by the participants despite the wide range of topics covered. These were: Social Anthropology and Ethology (Robin Fox), Social Anthropology and Psychology (Esther Goody), Social Anthropology and Development (Peter Lloyd), Social Anthropology and Oral and Literary Sources (Ravindra Jain), Structuralism (Nur Yalman), Marxist Analysis (Maurice Bloch), Mathematical Analysis (Clyde Mitchell), Transactions (Bruce Kapferer), Systems of Thought (Edwin Ardener), The Analysis of Symbolism (Roy Willis), and Problems of Fieldwork (Julian Pitt-Rivers). The Conference ended with general statements from the Chairman (Professor Meyer Fortes), Professor Raymond Firth, Professor R. F. Salisbury, and Dr Ralph Grillo. There were daily attendances of two hundred and fifty or more, including a large international group of guests and observers from, for example, France, Germany, the United States, Sweden, South America, Denmark, Holland, Israel, and Japan, as well as very many from the countries of Commonwealth tradition from which the Association draws its main membership. Most senior members in the British profession were able to attend part, at least, of the sessions, and many made contributions from the floor. In addition, it was frequently pointed out that there were present on this occasion more generations of British social anthropologists than at any Association meeting in the past: from recent graduates to founding members of the modern subject. The uniqueness of this occasion was underlined by the fact that this turned out, sadly, to be the last public appearance before his colleagues of the Life President, Professor Sir Edward Evans-Pritchard, who opened the sessions of the second day.

No one can make an easy assessment of the intellectual impact of this conference. A look back on the 1963 gathering may offer us some lessons. While that ASA conference does in retrospect seem to mark the close of one period of social anthropology, its 'new approaches' turned out to hint only remotely at the kinds of controversies that were actually to fill the next decade. All that can be safely said without adding fresh controversy is that controversy did become commonplace, while France, hardly mentioned by the 1963 conveners, frequently supplanted Britain or the United States as a source of new or fashionable theory.

On the whole, the topics and styles of the 1963 contributors did not necessarily or entirely prefigure the immediate future, although a number of them were important in it. The joint conveners wrote 'that the essays . . . reflect the feeling of their younger colleagues', and that they favoured 'clarification, the breaking down, and the refinement, of standard concepts, together with the closer specification of narrower social contexts . . .' That remained in part true, but the very process of clarification was changing the view that social anthropologists had of their subject, in ways that 1963 did not directly express. Things in fact moved more quickly, some thought more interestingly, others thought more hastily (many thought less surely) than they had been doing.

It is certainly doubtful then whether the 1973 Conference will turn out to be more predictive of 'new directions' than the last was of 'new approaches'. It should best be regarded as the record of a particular moment, with no exaggerated aims. In this respect the Conference was of its age. Its structure and its topics were approached perhaps with a little more reserve, even scepticism, than greeted 1963. For some the range of topics and their 'interdisciplinary' appearance seemed to represent a dispersal of academic energy. For some the balance appeared to have moved in the direction of various 'ideological' or 'transactional' analyses, and away from more tried methods. For others, on the contrary, the chosen topics were already ageing or old and did not reflect the most promising lines for the future. For still others the occasion appeared merely as a useful tribal rite of little intellectual significance. A women's session met amicably outside the official programme. Some radical leaflets were circulated. The third world now figured as a political as well as an academic subject. The historical period at least (it may well be thought in 1983) was unmistakable.

But, taking as neutral a view as possible, it is quite reasonable to say that, by contingency as well as by design, the volumes published in this series, which are based on a selection of the sessions, represent a number of subjects developed in social anthropology either relatively newly, or in greater depth, or with greater insight, over the last decade. This is certainly true of the first three titles: *Biosocial Anthropology, Marxist Analyses and Social Anthropology*, and *The Interpretation of Symbolism*; which will be followed shortly by *Text and Context, Transaction and Meaning, Numerical Techniques*, and *Structuralism*. All these titles represent recognized and authoritative developments in the current repertoire of social anthropology, which have grown from deep roots in the subject over more than one generation. Although not every social anthropologist in this more specialized age will be competent in or even sympathetic to all of them (they incidentally straddle several theoretical divides), they reflect a number of characteristic concerns in the subject today which, taken together with the regular annual series of ASA publications, suggest that social anthropology still retains its traditional curiosity and adventurousness—a feature emphasized by the number of younger contributors represented. Even a 'Decennial Conference' can do no more than that.

The present volume is based on a session convened on 6 July 1973. Editors have been free to seek amendments or additions to the papers in the sessions they convened, and to shape their volumes as they saw fit.

The ASA would like to thank warmly the President and Fellows of St John's College, Oxford, and the Warden of Rhodes House, Oxford, for providing the home for the Conference, and the staff of both for their willing assistance. It records its deep gratitude to the Wenner-Gren and H. F. Guggenheim Foundations for their welcome financial help. The major convening task was shared with me by Shirley Ardener, who found smooth solutions to many problems. In addition, the practical arrangements would have been impossibly complicated without the help of Nigel Barley, Martin Cantor, Christine Cooper, Robert Heath, Joy Hendry, John Mathias, David Price, Matt Schaffer, and Drid Williams, · all then mostly graduate students of the Oxford University Institute of Social Anthropology.

As General Editor, I should like to acknowledge the role of the present chairman of the ASA, Dr Jean La Fontaine, and her Committee, in establishing this series.

E. W. A.   © Edwin Ardener 1975

Robin Fox

Introduction

When behaviour implies elements acquired through experience, they are acquired according to a *program*, and that program is innate—that is to say, genetically determined. The program's structure initiates and guides early learning, which will follow a certain preestablished pattern defined in the species' genetic patrimony. Thus, in all likelihood, is the process to be understood whereby the child acquires language. And there is no reason not to suppose that the same holds true for the fundamental categories of cognition in man, and perhaps also for a good many other elements of human behaviour, less basic but of great consequence in the shaping of the individual and society.

Jacques Monod

The science of man, if it is a science, is one of the most unsettling and at the same time most curious lines presented for our investigation.

Lord Melbourne

The papers in this volume are intended to be representative of the biosocial approach to some persistent problems of anthropology. This is sometimes called the 'ethological' approach because of the obvious influence of the work of, for instance, Lorenz and Tinbergen, on the study of behaviour. The organizers of the conference in fact dubbed the session at which the papers were given 'Ethology and Social Anthropology'. But this is not an altogether helpful designation since on some areas of the biosocial approach ethology as such has had relatively little influence, and even in those areas most sensitive to its influence it is but one force among many. 'Evolutionary', 'neo-Darwinian', 'biological' are again variously used to describe this heterogeneous movement. There is perhaps no one exact adjective and all the foregoing are more or less correct, but I have chosen biosocial on the analogy with biochemical, intending to convey a concern with the biological factors in social existence; a concern which embraces social behaviour, social processes, and social institutions, and hence, to push the analogy to its conclusion, is a true 'biosociology'. I prefer this to the established 'social biology' which connotes eugenics and demography in a rather narrow sense. But because sociology as a discipline has been very little concerned with evolution and with man's physical nature while anthro-

1

pology has always maintained an interest in these, I will use 'biosocial anthropology'. Again 'bioanthropology' has been used, but is somewhat superfluous since anthropology is basically a biological science. By using 'biosocial' I hope to stress that this approach is indeed concerned with social man and social behaviour, and is an injection of biological thinking into social anthropology.

Biosocial anthropology is strictly a branch of evolutionary biology—as are ethology, population genetics, physical anthropology, etc. It is 'neo-Darwinian' and 'evolutionary' in that, within the framework of the neo-Darwinian synthesis, it accepts as its premiss the role of mutation and natural selection as the major factors determining the evolution of form and function in all species including our own. And it accepts this as the main point of departure for the analysis of anything concerning the life processes of any species. It views social behaviour, then, as the outcome of an evolutionary process and analyses it as such; culture itself being an outcome of the same process and only understandable in these terms. It incorporates the data of comparative sociology (social anthropology), but analyses them in a different framework of assumptions from those of the prevailing 'superorganic' and Durkheimian positions.

This is not the place for a detailed history of the development of this approach. It owes much to the new synthesis of Julian Huxley and his colleagues; to the breaking of the genetic code; to the renaissance of primatology under Washburn; to the upsurge of information on human evolution; to the rise and maturity of ethology as a science; to the revolution in linguistics by Chomsky; to the emergence of structural anthropology under Lévi-Strauss; to developments in neuro-science and endocrinology; and, negatively, to a growing dissatisfaction with the behaviourist–culturalist–environmentalist bias of the modern social and behavioural sciences. It is not unrelated, as an intellectual movement, to the rise of a new humanism which is perhaps as much 'neo-Kantian' as we are neo-Darwinian. That the whole business should appear not only mystifying but even threatening to sturdy functionalist–empiricist social anthropologists is not surprising. One can only hope they will eventually overcome their fright at discussions of the 'mind' on the one hand and the 'innate' on the other; in the meantime we must press ahead.

The disciplinary areas contributing most to the biosocial approach are:

1 Comparative sociology I am thinking in terms of a wide and generous interpretation of 'sociology' to include all the social sciences concerned with the natural behaviour of man in society. The data of history, mythology, linguistics, social psychology, human geography, psychiatry, ecology, etc., will be variously relevant. The criterion of relevance is not disciplinary. Any material might be relevant as long as it throws

2

light on species-specific patterns of adaptedness and their range of variation.

2 Comparative zoology   By this I mean the comparative study of non-human animal behavioural and social systems. Strictly, of course, this should include the above, and in an ideal world it might. But for the moment we must regard it as a separate disciplinary area on which to draw. It includes ethology proper, but data derived from other than ethological methods can be valuable. Primatology would come under this heading, and has a very special place in the overall approach.

3 Physical anthropology   I am thinking particularly of the study of human evolution, the fossil record, palaeo-ecology, and the dynamics of evolutionary change, evolutionary genetics, etc. This again is a generous interpretation of 'physical anthropology' that would equate it with an evolutionary biology particularly concerned with human phylogeny.

4 Primate biology   Again, I am not thinking in terms of a narrow conception but of the study of human and non-human primate life processes in a genetic framework—with endocrinology, neuro-science, the life-cycle, circadian rhythms, etc.

I have called these 'disciplinary areas' rather than disciplines, since they do not correspond with established boundaries, and the work of some disciplines will obviously straddle several areas. Also, not all the data from all of the above areas will be relevant to the biosocial approach (a good deal of physiology, for example)—only those that are concerned with social behaviour *per se*.

Biosocial analysis, it should be emphasized, is an analysis of the interplay between biological 'givens'—whatever their nature—and cultural responses. Many responses are possible, but always to the same givens. What this mode of analysis opens up is a new possibility for getting at the givens and hence a better chance to understand the variation in the responses. The *relationship*, however, has to be treated as problematical, and itself part of the life process, as Monod envisages (see quotation above).

Certain points of departure are emerging as useful in such an approach. Foremost is a *concern with the life-cycle*. This is, in a sense, the fundamental given and the one where the unfolding 'program' is most easily identified. It is therefore the peg on which to hang at least preliminary analyses. Associated with this are two concepts much debated but useful as guides to inquiry: *ease of learning* and *critical periods*. That members of a species will learn easily those behaviours that have been most useful in its history of adaptation, is not only true, but much more adequate than the older 'instinct' formula. Indeed, under this rubric, one could consider instinct proper as a special case of highly-programmed rapid 'learning'. The ease with which men learn rules, language, aggression, etc., gives us clues where to start in analysing human patterns of adaptedness. On the second point—critical or sensitive periods—it is obvious that the organism is not only programmed to learn some *things* rather than others, but is equally programmed to learn them at some *times* rather than others. Here again, for example, 'imprinting'—certainly an 'instinctive' mechanism—can be seen as a limiting case of critical-period learning. But in humans (and other primates where it applies) such behaviour complexes as incest avoidance, affection, language, mothering, and even skills (in the 'latency period'?) seem subject to similar sensitive-learning episodes.

This leads to another departure point—*pathologies*. Where we see seeming distortions of behaviour leading to a breakdown of functions essential to the successful completion of the life-cycle, we have often important clues to the nature of 'normal' development. The pathologies of motherhood and sexual behaviour attendant on maternal depriva- tion are a well-known example. Seeming pathologies may offer other kinds of clues however: it may well be that 'cheating'—a behaviour undeniably at the centre of much human activity to do with social control—is a predictable and inevitable given if Hamilton is correct. A further point of emphasis is the *characteristic bonds* associated with the life-cycle. These are more likely to have a heavy biological loading than those not so associated, since they are concerned with the basic pro- cesses of reproduction and rearing. Also, in dealing with bonds, *com- parison with other primates* is important, since bonds which are general among primates (male–male for example) are at least *prima facie* likely to be important in human society. This applies to other significant relationships and processes; for example dominance/submission, peri- pheralization of young males, equilibration, attention. Along with all these considerations we should look at any data regarding *physiological processes* accompanying any of the above; for example, the relationship between testosterone levels, male growth at adolescence, facilitation of aggression, and maintenance of dominance status.

Finally, for the student of social institutions, perhaps the most impor- tant point of departure will be *universals*—either institutions or pro- cesses. I prefer to look for the latter and to see this as a search for variables rather than fixed elements. Thus, in my own essay here, I do not look for institutions in kinship systems which crop up in every society ('nuclear families' for example)—an unlikely prospect to say the least—but for variables common to all kinship systems, however differ- ently they are expressed or utilized in the different types. The universal features we should look for, therefore, will more nearly resemble the elements of Chomsky's universal grammar: not fixed items that crop up in every language, but processes to which all languages must conform to count as human languages at all. The particular natural languages— that is, on our analogy, particular societies and cultures—would be predictably very different. The point about variables, after all, is that they vary. Thus, with such a variable as aggression, we would expect some few societies to have very little institutionalized or spontaneous aggression, while some others would have an extremely high level, and most would fall somewhere in between on a more or less normal curve. It is when we come to look at the interplay of these universal variables, the one with the other, in the context of particular cultural histories, that the real sociological part of the biosocial analysis begins.

These are, as I have said, simply some points of departure, they are not methods of analysis. There is no one method of analysis and an anthropologist interested in comparative kinship systems will obviously proceed differently from a developmental psychologist concerned with universals in child behaviour. But for a complete picture of 'biosocial man' to emerge, the data of the four disciplinary areas must be used, and the points of departure mentioned here must be covered. It will, however, take many partial analyses—such as those in this volume— to build up such a picture. The brief of the anthropologist, it would seem to me, is to pull together the various analyses in the framework of human evolution on the one hand, and the comparative study of society on the other. Within the former the enormous importance of the *hunting transition* should be given due weight, while the latter provides us with the *range of variation* open to human social arrangements; the end-product, as it were, of behavioural evolution in our species.

What I have described above is not a plan of action that any group of scientists is consciously following, Indeed, what is happening is a kind of convergent evolution, and I was fascinated to be able to bring together at Oxford such a seemingly diverse group of people and find that there was perfect ease of communication among them, given the 'neo-Darwinian' orientation which they all shared. Whether there was any communication between them and the rest of the assembly is an open question. There was a remarkable convergence of interests with the session immediately following ours on 'Psychology and Social Anthropology' which concentrated entirely on the search for universals in human cognition and perception. One might have hoped that the 'Marxists' who were there in fashionable numbers would have shown as much interest in Darwinian analysis as did Marx himself; and that the 'structuralists' (or is it 'post-structuralists'?) would have seen that their starting-point and ours—the nature/culture divide—was the same. A materialist science of man that explores innate properties of mind should encompass both these movements—but perhaps that is being imperialistic.

Not all aspects of the approach were covered, by any means, since we were limited to six contributors. I particularly missed a contribution from physical anthropology proper, and would have liked, say, a paper from John Napier on the evolution of locomotor functions and their relation to behaviour. Tiger nobly undertook to summarize the extensive and difficult material on somatic states and behaviour. He approaches this as a social scientist concerned to see where evolution has left its more obvious marks on behavioural processes, and therefore is more

5

likely to see the relevance of much of the data than, say, an endocrinologist. The non-anthropologists (Bischof, Chance, Blurton Jones, and Hamilton) have all interested themselves in anthropological questions, and I asked them for this reason to join us at the conference.

I have arranged the papers in what I assume to be a descending order of familiarity to social anthropologists. My own is deliberately, perhaps even provocatively, directed to a central concern of social anthropology, the nature of kinship systems. It illustrates one use of primatological data in analysis, and I have been fairly lavish with this to demonstrate its richness and suggestiveness. Since Bischof not only deals with the incest taboo—another central topic—but has made a magnificent attempt to come to grips with the anthropological formulation of the problem, I put him next. This is also the obvious place for him, since his conclusions are so close to mine and totally independently arrived at: that we must be cautious before declaring with certainty that all human social rules are 'intrusive' into nature; rules and taboos may simply be ways of coping with what would occur naturally anyway. It is rare to find a natural scientist paying such serious attention to the problems of social science—would that the social scientists might reciprocate. His essay also illustrates the importance of comparative zoology, although as a scientist he has largely been concerned with biocybernetical analysis of perceptual and motivational systems in animals and men. Blurton Jones is an ethologist of the classical school who has turned to the study of human children, not only 'ethologically' but cross-culturally. His essay illustrates not only the application of ethological methods to human behaviour, but also the relevance of this for a concern with universals and, more radically, a 'physical anthropology of behaviour'.

Chance has long been in communication with anthropologists and social scientists, and has served as a member of the council of the Royal Anthropological Institute. His work on monkeys both in the field and the laboratory has led him to formulate such variables as 'attention structure' which are directly applicable to the dynamics of human groups. Indeed, he has been carrying out research on human groups for some time. In the present essay he refines his concepts to take account of the difference between attention gained by threat and that activated by display, and introduces the new idea of 'advertence'. More importantly, he relates these two forms of cohesion in society to the development of intelligence in the individual. Tiger, basically a political sociologist, is concerned with the point of departure I mentioned above as 'physiological processes'. As we have seen, he approaches these not as a physiologist, but as a social scientist intent on discovering the internal mechanisms that correspond to external social processes. That many of these—for example, the synchronization of menstrual cycles among closely associated women—have all the characteristics of 'social facts', should give cause for contemplation to unreconstructed Durkheimians. Finally, Hamilton brings the findings of mathematical genetics to bear on the evolution of social behaviours such as altruism and cheating. It has been a constant puzzle to students of evolution that altruistic acts would seem to reduce the reproductive fitness of the altruist, and hence be inexplicable except by recourse to group selection arguments. Hamilton has demonstrated that if we consider the individual's genes, rather than his person, this is not so. His concept of 'inclusive fitness' represents a considerable breakthrough in our ideas about the evolution of behaviour, and its implication for an anthropological understanding of the nature of kinship as a part of the natural order is profound.

Some readers may find, as I did, the mathematical argument heavy going. They should perhaps try to grasp quickly the implications of it and pass on to the verbal argument. They will be rewarded by an astonishingly rich discussion of migration, xenophobia, warfare, language, intelligence, selfishness, cooperation, spite, cheating, and justice as interrelated phenomena in the evolution of human social behaviour. Nothing human is alien to the science of biosocial man, and Hamilton's essay illustrates vividly how important it is to start looking at all such phenomena as part of the evolved behavioural repertoire—as part of our species pattern of adaptedness, if we are fully to understand them. They are part of what Monod calls 'the species' genetic patrimony'; they involve experience for their full realization, but experience 'acquired according to a program'. It is to the unravelling of this program that the contributors (and many of their colleagues) have dedicated themselves—and in turn they have dedicated the volume to two great unravellers who have made the task that much easier.

The ultimate message in all this is a response to the ultimate question of social anthropology and social philosophy: the question of the possibility of social order. And the response is that of Aristotle rather than Hobbes: that the nature of order is part of the order of nature. It is not that man is as culture does, but that culture does as man is. Thus perhaps we can return, by the biosocial route, to the only question that allows anthropology (social or otherwise) a distinct existence among the sciences: what is man?

Note

To document all the points made in this introduction would require a bibliography of several hundred items. Most of the relevant literature up to 1971 can be found in the extensive bibliography of *The Imperial Animal*, by Lionel Tiger and myself (New York: Holt, Rinehart and Winston; London: Secker and Warburg). I would like to thank all the contributors for their help in presenting the papers to the conference and in preparing this volume; Edwin and Shirley Ardener and all the students and staff at Oxford for their helpfulness; Juliet Blair for editorial assistance; Maurice Freedman and the Institute of Social Anthropology at Oxford, whose hospitality I enjoyed while preparing the book; and finally the Harry Frank Guggenheim Foundation and its president Mason W. Gross, who made it all possible.

Robin Fox

Primate Kin and Human Kinship

INTRODUCTION

One of the main differences between the biosocial approach in social anthropology, and what might be characterized as the superorganic approach, is that scholars interested in the phylogeny of behaviour have to take as problematical the relation of cultural to natural factors. Those who espouse the dominant superorganic theory treat 'culture' as a self-evident category of 'non-genetic' behaviour: mostly, they see culture as 'intrusive' into 'nature'. Thus, the incest taboo, or kinship systems themselves, are seen as rule-governed and, hence, free products of the human imagination which are imposed on 'natural' tendencies. In the extreme form of this doctrine human behaviour is 99 per cent 'cultural' whereas animal behaviour is 99 per cent 'genetic'. Thus, animal breeding systems are genetic, while human kinship systems are cultural, and so on. The biosocial approach does not deny that culture can intrude into or impose itself upon nature, but (a) it regards this as a problem not a dogma, and seeks to establish of any widespread human institutional or behavioural pattern whether or not it is intrusive or simply an amplification or elaboration of 'natural' tendencies; (b) in its most extreme form (e.g. in Tiger and Fox 1971) it regards the issue as phony since 'culture' is simply the species-specific behaviour of a particular primate, and must be explained on the same principles as the evolved behaviour of any other primate. Here we will take the 'weak' view and simply ask to what extent human kinship systems are cultural intrusions and to what extent they are a natural product of the evolved behavioural propensities of the particular primate species *Homo sapiens* (or even of the genus *Homo*, since it is likely that *H. erectus* had them— at least there is no good reason why it should not). A full demonstration of the thesis proposed here would take us into subjects like the evolution of cranial capacity and its relation to sexual selection which I have dealt with at length elsewhere (Fox 1967a; 1972). Here I will be concerned with presenting evidence for a simple point: that our primate cousins have 'kinship systems' which contain the elements of human kinship systems, but that no other primate combines elements in the way that we do. In other words, there is little in our systems that is not found in 'nature' (and, therefore, there is no reason to suppose that it

9

is 'unnatural' to us), but nowhere in nature is our system found in its entirety. The elements are common: the combination is unique. My contention is, therefore, that it is to the combination of elements that we must look for clues to the uniqueness of human systems, not to the elements themselves.

This is not the same as the approach that says humans have *rules* and this is what makes the difference. I take for granted that we have rules since we are an animal that makes rules; the real question is do the rules represent more than a 'labelling' procedure for behaviour that would occur anyway? It seems to be Bischof's conclusion on the incest taboo that the taboos and rules indeed label 'natural' tendencies, and also that while all these tendencies exist in nature the human combination is unique. My conclusions about kinship, therefore, match his about incest. Of course, in the same way, there will be a variety of different styles and content to rules, varying with the cultural experience, ecology, and history of different peoples. (Much the same is true of other primates, as it happens.) But the contention here is that even in the absence of cultural rules and the logic of human imagination there would be kinship systems anyway, and that much of the rule-making and imaginative logic is simply (or complexly) playing games with a quite elaborate raw material. To those for whom kinship is a matter of categories this must be mystifying. But categories simply mean that linguistic labelling is possible, and that hence, with communication operating in another dimension than gesture and face-to-face contact, greater elaboration of the basic system is possible. In the absence of language, if a group of animals shows a tendency to treat certain classes of relatives differently from others, then it has invented categories even if these are not linguistically labelled. Discriminatory behaviour creates 'category systems' which, while more limited than linguistic systems, operate in much the same way. Thus, in a linguistic system I might act in a specific way towards a relative because he is recognized by his label; in a non-linguistic system I act towards him in a specific way because he is recognized by some other mark of distinction: his proximity to, or preferential treatment by, 'mother' for example.

Without labouring this point further, I will state that the problem here is to ask (a) what are the characteristics of kin relationships among primates; (b) how do these differ from the most rudimentary of human kinship systems; (c) wherein lies the crucial difference?

Previous attempts to answer this last question have concentrated on such things as (a) development of a 'pair bond' in man; (b) origin of 'families' in man and particularly 'nuclear families'; (c) the repression of unregulated sex in favour of regularized mating; (d) the reduction of hormonal control over behaviour permitting non-seasonal mating—etc. These probably all miss the point. They do not take into account either the complexity of primate kinship systems, or the true differentiating feature—exchange of mates.

10

Older theorists speculated on the 'earliest conditions of man', and as we know debates raged between proponents of 'primitive promiscuity' and 'primitive monogamy'. The former was usually seen as a prelude to 'matriarchy' (now popular again) and the latter to 'patriarchy'. This has all been dismissed as ridiculous for well-known reasons. But I think we can now go back to the question in a different way. We know a great deal about primates which can tell us what is behaviourally available to our order in general and, therefore, what must have been available by way of a behavioural repertoire to our ancestors. We also know that there must have been a gradual transition—'hominization' process—in which cultural and somatic features evolved together and in a feedback interaction. 'Early man' then, in this sense, was less like modern man gone wild than like a primate tamed. And even if we cannot deduce accurately the kinship systems of early man from those of the most primitive humans, we can do something better—we can distil the essence of kinship systems on the basis of comparative knowledge and find the elements of such systems that are logically, and hence in all probability chronologically, the 'elementary forms of kinship'. How we put all this information on primates, hominid evolution, and human kinship together is another question; for the moment let us establish this elementary pattern and work from there.

I will simply make a bald assertion clearly derived in part from Lévi-Strauss but also obvious in all my own work: the two elementary functions of human kinship systems are what I will call, borrowing somewhat recklessly from the jargon of social anthropology, descent and alliance. Descent defines who belongs to what category or how any category is related to any other category. In other words, kin are 'grouped' and the grouping of people in this way, the deciding of who belongs with whom, etc., I am calling descent for convenience. Alliance refers to the allocation of mates. In all human kinship systems people have assigned mates, and the system determines who can mate with whom—crudely who gets whom, who is allied with whom. Whatever else, in other words, a kinship system may or may not do, it has to do these two things: establish who belongs to what group (descent) and who can or must mate with whom (alliance). So, at its most basic, what I am saying is that if we had a heterosexual group of human beings, even without a language and cultural traditions, they would group themselves on a kinship basis, and they would allocate mates among themselves. They need not even do this according to 'rules' initially. All that we need assume is that they would find a 'means' to do this— that their behaviour would tend in this direction and that a social system would emerge on this basis.

My contention about non-human primate systems is simply this: both elements—descent and alliance—are present, but never in the same system. The uniqueness of the human system, therefore, lies not in inventing something new, but in the combination of these two ele-

Robin Fox

ments so that the mode of descent itself determines the nature of the allocation of mates: in other words, exogamy. I am by no means the first person to say that the essence of human kinship is exogamy—it is a great tradition, and I dedicate this essay to the memory of John Ferguson McLennan—but I may have found a new way of getting at the nature of this uniqueness.

TYPES OF SOCIAL SYSTEM

Very roughly we can distinguish old-world non-human primate social structures on the basis of the number of adult males in the group. The two 'types' that emerge are (a) one-male groups and (b) multi-male groups. I am including primates with 'nuclear families' in the one-male group category. The family of one adult male and one adult female with dependent offspring can be seen simply as a limiting case of the one-male group, which otherwise involves the 'polygynous' family with one adult male, several adult females, and again the dependent offspring. Species with nuclear family organization do not seem to combine these families into larger herds or bands (e.g. the gibbon and many prosimians). The nuclear family system seems basically territorial, with dispersion of dependent offspring. The polygynous or harem species, on the other hand, do often combine, and while the families remain distinct breeding units they unite into larger herds for protection, sleeping, etc. When this happens (as with e.g. hamadryas baboons, gelada, etc.) the herd structure includes not only polygynous families but the 'peripheralized' younger males who have yet to acquire harems of their own.

This total herd structure comes then to resemble in some ways the multi-male system (found in many species of baboon and macaque) where the 'families' do not exist, where there are several males in the group, and where the breeding system is on a 'consort' basis with brief breeding encounters between the more dominant males and oestrus females. Again the younger males are 'peripheralized', but instead of having to collect harems they have to work their way back into the 'centre' of the system so that they too may breed. The multi-male systems differ in the extent to which they are dominated by a central male who is their focus of attention (mostly the open-country species) or alternatively dominated by the older males generally without a 'principal breeder'. The difference is purely one of degree. Among savanna baboons, for example, there tends to be a dominant animal who can also dominate the breeding process by having first choice of females in oestrus; but in a large group there will always be several females in heat and several large males consorting with them. Among forest-dwelling chimpanzees—a multi-male group type—there is again a dominant male, but he is less likely to dominate breeding. But even here older and more dominant animals form consort relationships with

12

females in oestrus, and younger males are relatively peripheralized. The differences have been much overdrawn. Chimpanzees and common baboons, for example, are, in being 'multi-male' types, much more like each other than either is like the hamadryas or gelada baboons with their 'one-male' systems.

What all these systems have in common is a threefold division of the larger group into (a) adult males (b) females and young (c) peripheral males. We can look at any primate social system, including our own, in terms of the 'accommodations' made between these three blocks.

THE ONE-MALE GROUP

The one-male group has not had as much attention as its multi-male counterpart. The early work of Washburn and DeVore brought the latter into scientific and public attention, and most speculation about human origins rested heavily on the multi-male evidence. The work of Zuckerman (1932) on hamadryas baboons, although it showed clearly that a harem structure existed, was disparaged because it studied monkeys in 'concentration camp' conditions. The work of Goodall, Reynolds, and Schaller on chimpanzees and gorillas dismissed these from the one-male, harem-owning category. The first modern report (langurs were written about as early as 1902—see J. F. G. 1902) was by Kummer and Kurt in 1963 on the hamadryas (see also Kummer 1968). The patas monkey was reported on by Hall (1965) and the gelada baboon by Crook (1966) (see also Crook and Aldrich-Blake 1968). Indian langurs received attention from Sugiyama (1964, 1965a & b, 1967) (see also Sugiyama *et al.* 1965). Struhsaker's series of reports (1967a, b, c, d, e; 1969) on various *Cercopithecus* species, showed that while the vervet—the best known—had a multi-male system, many species (including of course *Cercopithecus patas*) had one-male groups. The colobus monkey (Marler 1969; Nishida 1972) appears to have this form of organization.

This structure was originally thought to be associated with the dry savanna and desert ecologies (Crook and Gartlan 1966), but the cerco-pithecine, langur, and colobus data suggest that forest species might also prefer it. However, there is a difference. The forest species tend to prefer relatively solitary families, while the savanna and desert species (with the exception of the patas monkey) prefer to encapsulate the harem group in a larger herd. Assuming the forest mode of existence to be more basic, it is easy to see how, when a species moved out into the open desert environment, it would amalgamate for protection. The patas monkey took the other route, and instead of safety in numbers sought security through dispersal and flight combined with diversionary tactics of the male—rather like some ground-nesting birds.

The data we now have show that a social system based on one-male groups is characteristic of the following species: *Cercopithecus ascanius*;

13

C. campbelli; *C. cephus*; *C. diana*; *C. erythrotis*; *C. l'hoesti*; *C. mitis*; *C. mona*; *C. nictitans*; *C. patas*; *C. pogonias*; *C. preussi*; *Colobus badius*; *Papio hamadryas*; *Theropithecus gelada; Presbytis johni*; *P. pileatus*; (*C. mitis*, Omar and DeVos 1971; on *P. johni*, Poirier 1970).

The basic unit is the polygynous family. This unit exists only as long as the male exists. The 'sons' of the male, at about one year of age, become peripheralized. How they acquire harems differs. Among hamadryas baboons, for example, an older male may take on an 'apprentice'—a younger male who comes to be tolerated more and more until he enters the group and eventually takes it over on the older male's death. Alternatively, the young males kidnap young females whom they 'mother' and who eventually become part of the harem. Either way, the 'father' of this avowedly patriarchal system is not in any sense succeeded by his 'sons'. There is a constant forming and breaking-up of the family units. Daughters may stay with the family but they may also be kidnapped into other families. If a male dies his mates are re-allocated. The only group that persists over time is the herd itself. And, in some species, e.g. the patas monkey, there is no herd. Among some langurs, groups of peripheralized males will invade a one-male group, drive out the patriarch, and fight among themselves. Eventually, one will emerge as the new leader of the harem (Sugiyama & Parathasanathy 1969). There are many variations. But the point I want to make is that although there is no enduring kinship group here, there is 'alliance' in my sense: a relatively permanent assignment of mates operates to produce distinct breeding units.

THE MULTI-MALE GROUP

Multi-male systems have been noted for the following: *Cercocebus albigena*; *Cercopithecus aethiops*; *C. sabaenus*; *C. talopin*: *Macaca fuscata*; *M. mulatta*; *M. silenus*; *M. radiata*; *M. uris*; *Papio anubis*; *P. cynocephalus*; *P. papio*; *P. ursinus*. In the late 1950s and early 1960s reports on multi-male systems came in from Bolwig (1959), Hall (1962a & b), Washburn and DeVore (1961), Maxim and Buettner-Janusch (1963), and Imanishi (1957; 1960). Throughout the fifties the Japanese workers had been publishing their data on macaques, largely in the journal *Primates*. The early volumes being in Japanese, the circulation was at first restricted. Altmann (1967) published a welcome set of translations, and subsequent volumes in English have continued the flow of data. Various compilations by DeVore (1965), Altmann (1967), and Jay (1968) among others, gathered together much of the material. Struhsaker (op. cit.) demonstrated this form of organization for the vervet monkeys (*Cercopithecus aethiops*). Chimpanzees and gorillas, as we have seen, have multi-male groups (Goodall, 1965, 1968, 1971; Reynolds 1965; Schaller 1963). Reynolds's notion of chimpanzee 'open groups'—contrasting markedly with the 'closed groups' of baboons, for example—rests on a confusion about the nature of the group (Reynolds 1966). Because the chimpanzee 'group' is spread out in the forest, with its bands of roving males and matrifocal units of females, this does not make it any the less a total group—a multi-male group in fact. Baboons in the forest do not have such 'closed' groups (Rowell 1966)—macaques never did anyway—while chimpanzees in open country have a group structure quite like the norm for multi-male savanna species (Itani and Suzuki 1967). This holds only for this dimension of social structure—kinship and the breeding system—and does not mean that there are not significant differences in other dimensions between, say, chimpanzees and baboons (see Chance, below).

The multi-male system contrasts with the one-male system in a rather startling way to produce a different form of social cohesion. The three 'interest groups' are the same, but the way they accommodate to each other is different. The adult males form a cohesive unit in direct contact with each other and arranged in a hierarchy. They are not divided from each other by their 'family responsibilities'. There are no distinct breeding units; the whole group is the breeding unit, with the rank-order to some extent determining mating preferences. In a sense, *all* the females in the group belong to *all* the males; consort relationships occur during the peak of oestrus, that is, during ovulation, but after that the female is returned to the pool. Recent work on chimpanzees—most of it still in progress—shows that they too have consort relationships; that high-ranking males may well get preference with females in oestrus; that older males restrict the mating behaviour of younger; and so on (personal communications from D. Hamburg, J. van Lawick-Goodall, D. Bygott, P. McGinnis, M. Thorndahl). It is possible that middle-ranking males may do better than those of highest rank. A really high-ranking male who is the centre of attention for his group may find it difficult to detach himself in order to go off with a female; his partners are not so handicapped. But older males as a body have first choice *vis-à-vis* younger.

Thus the mating relationship is quite the opposite of that in the one-male system. It is brief and non-exclusive. There is then, in my sense, no alliance: no relatively permanent assignment of mates. While this is not altogether primitive promiscuity, it could be construed as a kind of group marriage. However, the main point is that there is no alliance function at work.

There is, however, kinship. If the sexual relationship is brief and un-enduring, the consanguineal relationship is long lasting and of central importance. Earlier reports of this type concentrated heavily on the male hierarchy as the central feature, and it is indeed very important. But the existence of enduring kinship groups has now come to assume as great a significance. These groups are basically units of uterine kin—or not to put too fine a point on it, matrilineages. They were missed in earlier work simply because the longitudinal data were lacking. Now

15

Robin Fox

we know that in macaques and chimpanzees, for example, they are the enduring core of the group. It is the interplay between the male hierarchy, on the one hand, and the matrilines, on the other, that provides the dynamics of the social system.

The classic group consists of an old mother, her sons and daughters, and the children of her daughters. How long the continuity of a group will last we are not yet sure (after the old mother dies that is), but four-generation groups have been observed, and five or six are certainly possible. Sometimes, of course, we may be missing a matriline, since we have to infer that, e.g., two older females who are very close are 'sisters'—we do not know. But taking the typical three-generation group which is known, we find it in all multi-male systems of macaques and chimpanzees.

Studies of Japanese macaques (*Macaca fuscata*) and rhesus macaques (*Macaca mulatta*) have shown that members of matrilines act differently towards each other and towards non-members for many social purposes.

Yamada (1963) noted that related animals would tolerate greater proximity when feeding than would non-related, females being more kinship biased than males. He studied the effects of matriline membership on three categories of behaviour: (1) co-feeding (by which he means animals tolerating each other while feeding close together), (2) play, and (3) grooming. Animals which co-feed have a special tolerance relationship which permits contiguous feeding without tension. This relationship is almost exclusively found between members of the same matriline. The general rule is that frequent co-feeders are uterine kin while animals that are not blood relatives feed together rarely. In some cases, animals of equal social rank co-feed irrespective of kinship ties. Play is very seldom influenced by kinship and young animals play with whomever they can. Like feeding, grooming follows kinship lines in general but not so strictly. Co-feeding, he concludes, has to do with a 'family-making tendency' while playing, and to a lesser extent grooming, have to do with a 'group-making tendency'. Yamada also noted an interesting effect of rank: '. . . a higher individual has in general less tolerance of an individual not in kinship, while he gives his relatives special treatment.' In summary, and focusing on sex differences: 'With females, the higher the social status of the individual, the more frequently she co-fed with individuals in kinship and the more exclusively she behaved toward other individuals. Accordingly it may be supposed that the more prominent the status of the kinship group, the tighter its unity. With males, however, this fact does not always apply.' Males, for example, showed a smaller tendency to co-feed with maternal relatives.

Loy (1972), studying 33 juveniles separated from the rest of their group, could predict 95·6 per cent of dominance relationships solely on the basis of the known rank of the mother. Also, grooming between related monkeys was over five times more frequent than would be expected from a random selection of grooming partners. This is a more exact finding than Yamada's and shows that, while grooming is certainly 'group-making', grooming acts can cluster significantly among kin. Related animals were more likely to sit together, and on sitting to touch, than were unrelated animals. He concludes: 'These findings serve primarily to demonstrate the far-reaching effects of matrilineal relationships on rhesus behaviour and to reinforce the view that the matrilineal genealogy is the basis of rhesus monkey social organization.'

Kaufmann (1966) studied the maturation of infant macaques on Cayo Santiago. He found increasing interaction with members of the 'matrifocal family' as the young progressed from infancy through the juvenile period.

Sade (1965) notes that 79 per cent of touching or lying together occurred, in one sample, between matriline members. In a later paper (1967) he comments on rank:

> Offspring begin to fight as old infants or young yearlings. They defeat their age peers whose mothers rank below their own and are defeated by their age peers whose mothers rank above their own. The hierarchy thus established persists for several years, either because the first few fights set precedents which cannot easily be broken, or because the offspring continue to associate with their mothers or other older monkeys who rank near their mothers, or because of both reasons. As females become adult they come to rank just below mothers in the hierarchy of adults. This means they defeat not only their age mates but also older females who were adults when they were growing up. As males become adult they tend to rank near their mothers in the adult hierarchy, but at puberty or later, if they remain with the group, they may lose or gain rank. I speculate that at about puberty physiological differences between males become more important in fighting and that the difference that derives from past experience and continued association with adults of differing rank become less overriding in determining the winners of fights.

This finding is supported by Koford's data (1963).

Alexander and Roth (1971) looked at the effects of crowding on matrilines. The animals of the group were given progressively less space in which to move, and the rate of aggression, as we might expect, rose steadily. However:

> Crowding produced no significant effects on the behavioural preferences of family units. During crowding sessions the mothers and juveniles directed 64·8% of their affiliative behaviours and 6·6% of their aggressive behaviour toward uterine kin. During control sessions the corresponding figures were 66·6% and 4·6% respectively. These changes did not approach statistical significance.

Miller *et al.* (1973), on Cayo Santiago, studied various categories of

17

behaviour in 18 male macaques between 2 and 8, under the general headings of 'joining', 'grooming', and 'threatening'. They found that: 'Statistical analysis revealed that the male subjects engaged in more positive social interactions with the matrifocal family (mother and offspring) than with other members of the genealogy (matriline), and, in turn, more such behaviour was observed within the genealogy than with unrelated monkeys. Threatening behaviour within the genealogies was infrequent.' They conclude that the socialization of the males within their matrilines contrasts markedly in its non-agonistic nature with the socialization of the males into the non-kin group at large, where threat becomes most important. (To illustrate for social anthropologists the composition of some of these matrilines, I include in *Figure 1* six of those occurring in the Cayo Santiago Group E studied by Miller and his associates. The terminology of primatologists is somewhat loose, and 'matriline', 'matrilineage', 'family', 'matrifocal family', and 'genealogy' are sometimes synonymous, sometimes not.)

Vandenbergh (1967), studying captured monkeys released on to an island, found that the first groups to form were composed of related females and their offspring. These tended to be unstable unless they were joined by males and a typical 'multi-male' structure set up. However, relative to each other, they maintained a ranking system. 'The rank of a band seemed to depend on the vigour of the basic group of females more than on the dominant male, because even though the dominant males shifted from band to band the rank of the band remained unchanged.'

Kawamura (1958), in his classic paper, described the 'matriarchal' order of the Minoo-B troop. He distinguished basic rank—that determined by open competition between animals for food—from dependent rank—the rank derived from the mother. He observed that in the troop (in which the leader was, unusually, a female) the children rank, without exception, according to the rank of their mothers. Among brothers and sisters, the younger rank higher than the older presumably because of their closer association with the mother and the protected-threat status they derive from that association. He concludes:

> This troop has a rank system based on the matriarchal order. . . . The matriarchal order is very stable, and conclusive evidence has been collected to support the fact that this order not only appears when ranking is tested by giving food, but that the troop lives its daily life under this order. . . . Classes of this troop arose, based on lineages. These classes form a social organisation transcending generations and they decide the social future of new-born babies.

This conclusion has been supported by the studies already cited and by Imanishi (1960) and Itani *et al.* (1963).

'Classes based on lineages'—this, more than any other finding about individual status, is the crux of the material we are summarizing here.

Figure 1   Genealogies of Group E: Cayo Santiago

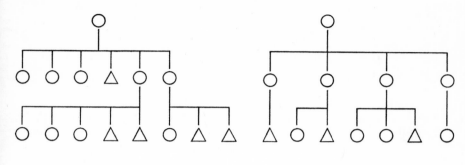
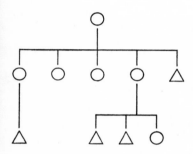
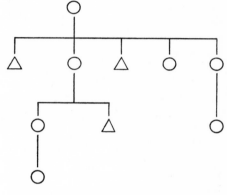
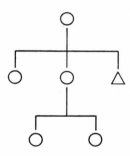
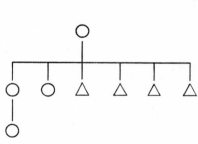

Source: Miller *et al*. 1973

Marsden (1968) followed this up with his observations on rhesus monkeys:

> With only one exception, all offspring directly reflected each change in the rank of their mothers, falling or rising to the position in the hierarchy occupied by the mother in relation to other family groups. More importantly, perhaps, patterns of aggression toward other adults and their young support the hypothesis that offspring often assist the mother directly in aggression toward other monkeys, and in turn are assisted by the mother. Therefore, it is more meaningful to think of families (mothers and offspring) as constituting social units which show or receive aggression from other family units in establishing a new hierarchy, or in maintaining an existing one. Thus a new concept of dominance based on family units is probably more realistic than a straight linearity of individuals when describing social dominance in rhesus bands.

If we add to this what is now known about the cohesion of expanded units of related mothers and offspring, that is, matrilineages, then a forceful picture emerges of the dynamics of a social system based on the interplay between ranked lineages, on the one hand, and the linear hierarchy of males, on the other.

Jane van Lawick-Goodall, in the several papers and books already cited, has described for chimpanzees the strong and enduring bonds between mothers and offspring and among the offspring themselves. The qualitative aspect of these relationships comes out clearly and graphically in her work (see also Goodall 1967). Rowell (1972) cites a communication from I. C. Kaufman to the effect that 'pigtail (macaque) social organization is based on strong permanent mother-infant ties leading to the formation of socially exclusive matrilines. . . .' (p. 149), thus adding to the growing list. Little as yet is known about the common baboons for the simple reason that we have no longitudinal studies comparable to those on the macaque. But given all the other similarities between such closely related animals, we would expect them to be similar in this area. Ransom and Rowell (1972) have provided suggestive data on this score.

Finally, we should look at the most extensive data so far on the Japanese macaque, that of Koyama (1967, 1970). I will deal with this in more detail later under the discussion of group fission; for now, it is enough to note that he found sixteen lineages ranked in order one through sixteen. He comments (1970) on dependent rank generally: '. . . concerning 54 females 3–10 years of age, rank among the same-age monkeys always corresponds to the rank of their mother and is the same as that of 1964 referred to in the first report. Accordingly, no change in rank among females of the same age occurred during these two years.' Among males 3–10 years old in March (n = 39) only two males had a rank higher than that of their consanguineal group. Rank

20

among age-mates was the same as the rank of their mothers. Older monkeys tended to rank higher, but in all exceptional cases the younger monkeys had mothers ranking higher than those of the older monkeys.

I have taken the trouble to lay out these data in some detail (although space limits the discussion of finer points) in order to establish the 'reality' of this kinship organization, an organization based on matrilineal ties between units of mothers and offspring. The cumulative evidence shows clearly that matrilineages exist; the lineages are ranked; behaviour between members of a lineage is significantly different from that with non-members; membership affects the rank of individuals; membership is very important in the socialization of males in particular. It is not simply that a high-ranking mother will ensure a high rank for her son; the matrilineages themselves being ranked, if lineage A ranks over lineage B, then all members of A rank over all members of B. This remains true over long periods for females and is initially true for males. Male rank may change later, but even then, membership in the lineage may be important, as when brothers help each other to rise in rank even when they change groups (Kaufmann 1967). If Chance's thesis concerning attention as the basis of rank-orders is correct, then, clearly, being of a high-ranking lineage is an important source of attention for males (Chance 1967; see the extension of this idea in Tiger and Fox 1971: ch. 2). The low level of aggression, on the one hand, and the performance of 'altruistic' acts for near kin, on the other, makes sense in terms of the theory of kin selection and inclusive fitness put forward by Hamilton (1963; 1964; 1972) and developed by Trivers (1971; 1972).

Thus the female-with-young block here is linked to the males not only by a breeding relationship, but also by uterine kinship. But although we have kinship, we do not yet quite have exogamy. The relationship between mating and lineage is uncertain. Sade (1968) originally proposed that mother–son mating was probably nonexistent. 'It appears that a male is inhibited from mating with a higher rank female only if she is his mother, since males will copulate with higher-ranking unrelated females. The inhibition is therefore specific to the parent–offspring relation. The role of infant reverberating in the relations of adult male towards his mother is apparently incompatible with the role of mate.' 'The rarity of mother–son mating is even more remarkable when the fact is considered that behaviour such as grooming, body contact and mutual defense occur frequently between mothers and some mature sons during the mating season as well as during the non-mating seasons.' Only 1·1 per cent of matings were between mother and son.

Missakian (n.d.), however, challenges these findings with her own data from Cayo Santiago. She had 59 possible mother–son pairs but 21 (36 per cent) were eliminated when sons left their natal groups; the emigrating males were all between three and five years old. Of 99 possible brother–sister pairs, 20 (20 per cent) were eliminated when the brothers left. This male exodus—noted by many other observers— obviously helped to limit inbreeding. The results she reports are based on 26 mother–son and 42 brother–sister pairs. 'Of the total of 243 instances of sexual behaviour observed, 14·8 per cent involved genealogically related animals and 10·6 per cent of the above figure involved either mother–son or brother–sister pairs. The combined measure of mount series plus copulations for mother–son mating is 5·4 per cent as compared to 1·1 per cent reported by Sade in 1968.' With one exception, all instances of mother–son mating involved males from three to five years of age—in fact 3–5-year-old males who stayed in their groups rarely mated at all except with their mothers.' Of the 26 pairs in which mating was possible, sexual activity was observed in 8 (31 per cent) of the cases. Among the 42 brother–sister pairs sexual activity was observed in only 6 (12 per cent) of the pairs. Again all the males involved were 3–5 years old. Dominance of the mother over the son or the sister over the brother did not seem to inhibit mating. Nor was incest limited to high- or low-ranking lineages. However, Missakian observes that 'normal' consort behaviour—following, genital examination, extensive reciprocal grooming, repeated mount series, and copulations—was not present in the mother–son matings.

I have spelled out Missakian's findings at length because Sade's earlier formulation has achieved almost the status of a law in some circles—even though he would be the first to adjust his conclusions in the face of further evidence. But we should note, as does Missakian herself, that the discrepancies are not as wide as might appear. For a start, her group was much larger and its dynamics may have been different from those of Sade's group. In a large group it is possible to see a greater variety of behaviours or the same behaviours more frequently. The relative level of sexual activity between mothers and sons is not much higher (5·4 per cent as opposed to 1·1 per cent) and the size of the group may account for this. And while 31 per cent of the mother-son pairs had some sexual activity, only 12 per cent of the brother-sister pairs did so. It is also clear that this activity is confined to the first breeding season of the male, and, interestingly, lacks the full complement of 'consort' behaviour. It is in the nature of a trial-run it seems. Several factors—adolescent sterility, inexperience, etc.—may render it unlikely that conception would result, and even if it did it would only represent a drop in the gene pool, as it were. It is important to get at the exact pattern of mating here, but at the same time, Sade's formulation is really only somewhat modified; his general proposition still stands, and the chances of mating occurring 'within the genealogy', as the primatologists say, are considerably lower than the chances of it occurring with animals not related by uterine ties.

We are here, of course, dealing with statistical tendencies not social rules, as was observed earlier; with statistical not mechanical models. What we have is a set of tendencies which make it probable that these

Figure 2 Multi-male group model

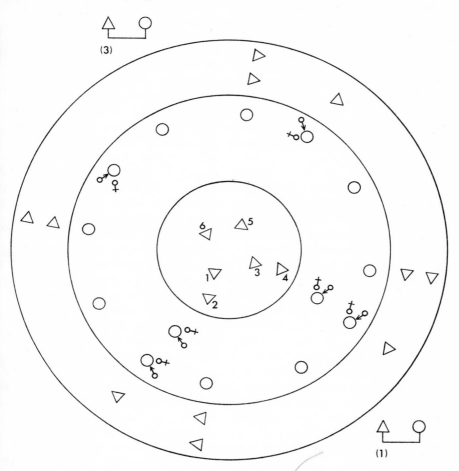

Inner circle: male hierarchy (1 = α etc.)
Intermediate circle: females, infants, juveniles
Outer circle: peripheral males
Outside: consort pairs

♂ ♀ infants still with mother

animals will display (a) patterns of grouping and behaviour which significantly discriminate between uterine kin and other animals, and (b) a statistical trend away from the probability of mating regularly with uterine kin over the 'breeding lifetime'. In the multi-male system, then, to use the jargon, we have *descent* but no *alliance*; kinship, but no marriage. There is assignment of kin to kinship groups, but not even a relatively permanent assignment of mates. The kinship system seems to have a discouraging effect on mating within the group—the beginnings of a negative marriage rule—but nothing more. There is thus a crucial difference between the two systems and between either and the human system, which combines both lasting kinship groups and polygynous and nuclear families; that is, to use the jargon again, both descent and alliance, kinship and marriage—and which uses the descent to determine, negatively or positively, the alliance.

MIXED SYSTEMS

There are clear differences between the genera in preferences for one or the other form of organization, suggesting that phylogeny, as well as ecology and ontogeny, is important. In *Cercopithecus* only three species are multi-male while twelve are one-male; in *Macaca* five are multi-male and none are one-male; *Papio* has four multi-male and one one-male. But we should also note that there are 'mixed' systems in which animals of the same species in different habitats gravitate towards one or other structure; for example, *Colobus guereza, Mandrillus leucophaeus, Nasilis larvatus, Presbytis cristatus, Presbytis entellus.* (On *C. guereza* see Marler 1969; *M. leucophaeus*, Gartlan 1970; *P. entellus*, Jay 1962; 1963a; 1963b; 1965.) Macaques at the limits of their ranges, and especially in difficult country, will tend to 'shed' males, for example, since these are not necessary for the breeding system and consume a great deal of food that could go to the mothers and young. There are several evolutionary possibilities. The 'basic' primate group could be the nuclear family which evolved into the one-male group and eventually into the multi-male group in the savanna; that is, by adding first more females and then more males. Alternatively, the multi-male group can be seen as basic and going in two directions: the shedding of males to produce the one male-group structure, or the shedding of both males and females to produce nuclear families. Various possibilities are discussed in Eisenberg *et al.* (1972), and Goss-Custard *et al.* (1972).

MODELS OF THE TWO SYSTEMS

In order to show graphically both the similarities and the differences between the two types, I have diagrammed them according to the same plan—the concentric circle of the 'classical' description. *Figure 2* shows the multi-male model as originally described for the common baboons.

Figure 3   Multi-male group model with kinship connections

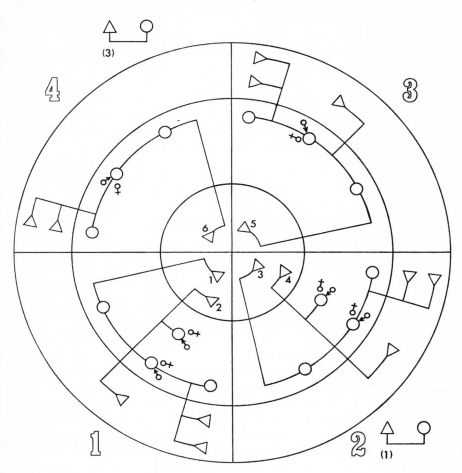

Inner circle: male hierarchy (1 = α etc.)
Intermediate circle: females, infants, juveniles
Outer circle: peripheral males
Outside: consort pairs
Quadrants: matrilineages ranked 1 through 4

♂♀ infants still with mother

25

Robin Fox

In the inner circle are the males of the hierarchy—one through six (the usual maximum number). In the intermediate circle are the females with and without offspring. In the outer circle are the peripheral males. Outside the circle—literally—are the consort pairs, and one must imagine various of the hierarchical males (here 1 and 3) temporarily out of the system for a few days at a time consorting with females in oestrus. The model is not totally accurate since in reality there would be more females per male. Before the information on lineages came in, this model would have served to illustrate discussion of the multi-male type.

Figure 3, which can be totally superimposed on *Figure 2*, shows the same system with the kinship ties added. I have envisaged four lineages ranked one through four, one in each quadrant. In the less senior lineages I have put fewer animals, but numbers and dominance are not necessarily connected. The rank of the males is seen to be determined by the rank of their respective lineages, although there will be exceptions to this with older males. Again, the consorting pairs are outside the system. The peripheral males in lineage 1 will be more likely than those in the other lineages to reach top status in the group, thus succeeding their maternal uncles and great-uncles.

Figure 4 is the one-male group model drawn to the same ground plan to show how the peripheries are linked to the centre. Again I have envisaged four groups. In the centre are the harem-owning males—unranked. Attached to them are their mates and their infants. At the peripheries are the young males, some of whom have kidnapped young females in order to start their own harems. In some of the one-male groups there are apprentices, waiting to take over the group on the death or retirement of the senior male.

What I hope this illustrates better than any verbal description, is the structural similarity that lies in the existence of the three 'blocks' as I have called them—senior males, females-with-young, and junior males, and the dissimilarity in the way in which the peripheries are linked to the centre; both in the way the females are attached to the males for breeding purposes and the way in which the junior males make it into the hierarchy. In the multi-male model the link is that of matrilineal kinship. Females are the link between the centre and the periphery in their role as the pivots of the kinship system. Males almost 'succeed' their maternal uncles and their brothers; and while they have to prove themselves, their birth status is an enormous help. Mating has no place in the cohesion of the system. Indeed it can be seen as virtually a divisive force and, almost symbolically, it has to take place physically outside the group. With the one-male system, on the other hand, the mating link is all that joins the centre to the periphery. Females are again the link but in their role as wives. There is no kinship link between the dominant males in the centre and the peripheral candidates, who must either kidnap their way back in or ingratiate themselves with an older

Figure 4   One-male group model

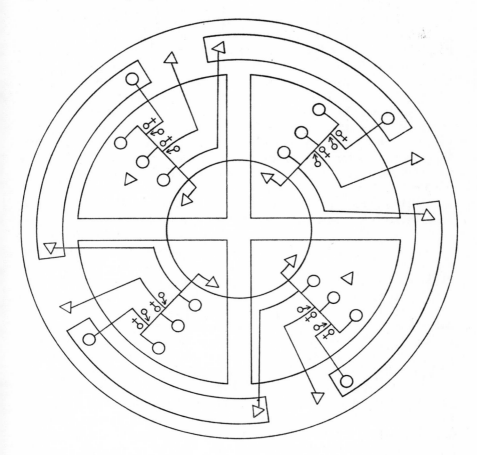

Inner circle: harem-owning males
Intermediate circle: females, infants, apprentices
Outer circle: peripheral males, kidnap pairs
Quadrants: one-male groups

♂♀ infants still with mother

27

harem chief. Kinship plays no part in the cohesion of this system, while mating is thoroughly institutionalized.

GROUP FISSION

Leaving aside for the moment the question why the evolving hominids put these systems together (if it can ever be answered), let us look more deeply into another feature of these multi-male, matrilineal systems: group fission. Primate groups split up from time to time when they get too large (a relative matter). We have one excellent account of such fission by Koyama (1967, 1970)—more are coming in (see Furuya 1969). In Koyama's group of macaques the lineages were ranked from 1 to 16. While they stayed together, each lineage would periodically feed new males into the status system and support them in their rise. Koyama suggests that as long as the rank order is stable all is well, but that if, for example, a high-ranking male habitually consorts with a low-ranking female this may cause a rise in the status of her offspring and hence upset the balance of the group. In any case, the group split, and lineages 1–7 formed group A, while lineages 8–16 formed group B. As Koyama points out, when a male leaves a group—and they often do— he leaves alone; when a female leaves she takes all her consanguines, and starts an exodus of those lineages ranking below hers. The tendency of the males to move about, however, has an interesting consequence. Most of the males from group A eventually ended up in group B and vice versa. Thus, we have a picture, in effect, of two 'related' groups, each consisting of ranked matrilineages, which exchanged males over a period. Had these males been involved in some system of alliance, in my terms, that is, had they ended up as relatively permanent mates in a matriline of the other group, then there would have been little difference, except at the symbolic level, between this and an 'Iroquoian' system of two moieties each composed of ranked matrilineages, and a rule of moiety exogamy plus matrilocal residence—all very human, and yet little more than a naming system away from the Japanese monkeys. If groups A and B were called 'Eaglehawk' and 'Crow', and the various lineages 'Snake', 'Beaver', 'Bear', 'Antelope', etc., then a picture emerges of a proto-society on a clan–moiety basis which would have delighted Morgan and McLennan (and Bachofen), but depressed Westermarck and Maine for sure. For one thing, it would have completely bypassed the nuclear family, and would not even require an incest taboo, much less a 'pair bond'.

We do not need to rehash the matriarchy–patriarchy debate here. But while there is some support given to Bachofen's contention that matriarchy is based on deep biological principles, we have to allow that, for our one-male-group species, so is patriarchy. But this is not a matter of paternity, as earlier thinkers maintained. Patriarchal baboons are not interested in offspring *per se*—in fact they can be quite hostile to them. They are interested in obtaining and controlling mates. Maternity may well be the basis of descent, but paternity has little to do with alliance. If we marry the alliance principle as we find it in primates to the descent principle as we find it, we do not get patriarchy or patrilineal lineages; we get something more like sororal polygyny and a system of matrilineages with a possibility of lineage and even band exogamy. To put it brutally: females are interested in offspring, males are interested in females, at least at this stage of the game.

It would be of intense interest to find any species or population of non-human primates combining the two principles. This would mean that even more was attributable to 'nature' than has been envisaged here. The question would then be: on what basis was the combination made? And my own feeling is that it would not be on the basis of exogamy as we understand it in human kinship systems. That the descent system itself should determine the allocation of mates will probably remain the uniquely human distinction.

CONCLUSIONS

The position we are at is this. If we look at the range of non-human primate social systems we find several possibilities represented there: fairly stable, ranked matrilineages; a movement of males between groups—and in chimpanzees of females between groups; stable mating associations of a male and several females. As we have seen, the last is not found in conjunction with the others, but they all exist in the primate order, they all have good adaptive reasons for existing, and they represent the raw material out of which primate kinship systems, including our own, are woven. The question for us to explore in the future is why the hominids put together the alliance and descent functions when no other primates had found this necessary, and what form this putting-together took. For various reasons it is unlikely that it could have been a pattern of dispersed nuclear families. This leaves the possibility of some kind of system such as that envisaged above, with males being attached to multiple unrelated females for mating purposes, and yet remaining attached to related females for 'kinship' purposes. In the male–male relationships, again, there would be one relationship to consanguineous males—in the matrilines—and another to unrelated males in the hierarchy. Patterns of elementary exchange between lineages on the one hand, and moieties on the other (originating in simple band-splitting), are easy to envisage—the monkeys are already almost doing it anyway.

The 'why?' question I have tried to answer elsewhere (Fox 1972). It is clearly to do with the severe ecological changes consequent upon the adoption of hunting on the open savannas. I have here been more concerned with the 'how?' question; the question of the raw material that would have been available to our ancestors in making the change.

If this gives credence to the matriarchal theory—that earliest kinship was matrilineal, that earliest sex was promiscuous, and that polygyny came next and eventually that patrilineality was 'invented' by the men —then so be it. I would rather see human kinship systems as symbolic methods of working out the tension inherent in trying to marry the polygynous (alliance) principle with its male emphasis, and the matrilineal (descent) principle with its focus on the female. Clearly, in the one the males are well in control; in the other the females are at least covertly exercising great influence. But in any case, to return to the original point, the dynamics of human kinship systems may well lie as much in deep and inherited biological tendencies as in the logic of human imagination—perhaps even more so.

The more general point being made here is that kinship groups and the alliances between them are not merely matters of rules, categories, laws, prescriptions, etc. They are more than results of the free play of human imagination. They are embedded in natural processes such as those described by Hamilton and Trivers. Relatedness—the sharing of common genes—is crucial to the process of natural selection, to the determination of what genes will dominate in any breeding population and ultimately what features will characterize the form and behaviour of an evolving species. Thus kinship groups are the outcome of natural processes; they are as natural as limbs and digestion; they are the outcome of adaptive responses and natural selection over millions of years; they are not peculiar to human society. They do not depend for their existence on the equally natural ability to classify and name which characterizes our species; in the absence of language and rules, they would still occur. It is only their elaboration that depends on the classificatory propensity and the evolution of true language. And even this elaboration itself, stemming from ecological circumstances, historical contingencies, and the vagaries of human inventiveness, is only a variation upon themes. Anthropology has been obsessed with the variations; I am suggesting a new look at the themes—not by a deduction of abstract 'functions' but by careful assessment of what 'nature' presents us with in conjunction with what we know about the selection pressures operating on our own species. Without a knowledge of the themes, we will never properly understand the variations—although we can get pretty close almost by accident.

Thus when Lévi-Strauss (1963) says that

is the 'atom of kinship', he is more or less right, although it might have been better diagrammed:

He is also right when he sees this as a pattern to which, under stress, societies revert. But the question is, why is he right? His basic point is that this follows from the incest taboo—the culture–nature divide— which forbids ○⌐△ as a mating relationship and so forces the △ = ○⌐△ situation. This 'builds in' the mother's brother/sister's son relationship. But we have seen that the latter relationship already exists in the multi-male system; also, the one-male group system has the △ = ○ formula. Both elements of his equation, then, exist in nature, but not in the same system together. Putting them together may or may not have required an incest taboo *per se*. The taboo may be simply the result of the evolution of inhibition I have described (Fox 1972). That same evolution, however, as I have argued, both facilitated and required the exchange of spouses, hence exogamy—a factor independent of who has sex with whom (a point made in Fox 1967b, to which Leach called attention in his critique of Lévi-Strauss; see Leach 1970). The relatively permanent mating bond already existed in nature—our 'alliance'; the permanent kinship group already existed in nature—our 'descent'. It only remained for the two to be put together by the evolving hominids.

The possible ways of doing this nowhere *require* an incest taboo, a nuclear family, or the faculty of speech—those much-touted 'origins' of truly human kinship. Logically it would have been much simpler for the hominids to move from primate kinship to human kinship by going directly to the most elementary of elementary kinship systems described by Lévi-Strauss (1969)—and this without any invention, without any intrusion of culture into nature, but with all the possibilities that speech added to the raw material so richly present.

NOTE

It will be clear from the many references to it that this article should be read as an extension of my 'Alliance and Constraint: Sexual Selection in the Evolution of Human Kinship Systems'. It is also a continuation of the themes discussed by Imanishi, Reynolds, Sahlins, Powell, *et al.* Space precludes any discussion of these here, but it will be obvious where I agree and where I differ. I would like to acknowledge the research assistance of Anthony Pfeiffer and Richard Diener. So much have they contributed both of information and ideas, that they should properly be co-authors, except that I am solely responsible for the final formulation of the argument from which they might well wish to dissent. I was enabled to write the article as a result of leave granted by Rutgers University, and the generous support of the Harry Frank Guggenheim Foundation, New York, NY.

Bibliography

ALEXANDER, B. K., & ROTH, E. M. 1971. The effects of crowding on aggressive behaviour of Japanese monkeys. *Behaviour* 39: 73–90.
ALTMANN, S. A. ed. 1965. *Japanese Monkeys: a collection of translations.* Atlanta: Published by the editor.

Robin Fox

ALTMANN, S. A. 1967. *Social Communication among Primates* Chicago: U. of Chicago Press.

BOLWIG, N. 1959. Study of the behaviour of the chacma baboon, *Papio ursinus. Behaviour* 14: 136–163.

CHANCE, M. R. A. 1967. Attention structure as the basis of primate rank orders. *Man* n.s. 2: 503–518.

CROOK, J. H. 1966. Gelada baboon herd structure and movement. *Symp. Zool. Soc. Lond.* 18: 237–258.

CROOK, J. H., & ALDRICH-BLAKE, P. 1968. Ecological and behavioural contrasts between sympatric ground-dwelling primates in Ethiopia. *Folia primat.* 8: 192–227.

CROOK, J. H., & GARTLAN, J. S. 1966. Evolution of primate societies. *Nature Lond.* 210: 1200–3.

DEVORE, I., ed. 1965. *Primate Behavior: Field Studies of Monkeys and Apes.* N.Y.: Holt, Rinehart and Winston.

EISENBERG, J., MUCKENHIRN, N. A., & RUDRAN, R. 1972. The relation between ecology and social structure in primates. *Science, NY* 176: 863–874.

FOX, R. 1967a. In the beginning: Aspects of hominid behavioural evolution. *Man* n.s. 2: 415–433.

—— 1967b. *Kinship and Marriage: An anthropological perspective.* Harmondsworth and Baltimore: Penguin.

—— 1972. Alliance and constraint: sexual selection in the evolution of human kinship systems. In : B. Campbell (ed.), *Sexual Selection and the Descent of Man 1871–1971.* Chicago: Aldine.

FURUYA, Y. 1969. On the fission of troups of Japanese monkeys. II. General view of troup fission of Japanese monkeys. *Primates* 10: 47–69.

G., J. F. 1902. Habits of the lungoor monkey. *J. Bombay Nat. Hist. Soc.* 14: 149–151.

GARTLAN, J. S. 1970. Preliminary notes on the ecology and behaviour of the drill, *Mandrillus leucophaeus,* Ritgen, 1823. In: J. R. and P. H. Napier (eds.), *Old World Monkeys.* N.Y.: Academic Press.

GOODALL, J. VAN LAWICK. 1965. Chimpanzees of the Gombe Stream Reserve. In: I. DeVore (ed.), *Primate Behavior.* N.Y.: Holt, Rinehart and Winston.

—— 1967. Mother–offspring relationships in free-ranging chimpanzees. In: D. Morris (ed.), *Primate Ethology.* London: Weidenfeld and Nicolson.

—— 1968. The behaviour of free-living chimpanzees in the Gombe Stream Reserve. *Animal Behaviour Monogr.* 1, part 3.

—— 1971. *In the Shadow of Man.* Boston: Houghton Mifflin.

GOSS-CUSTARD, J. D., DUNBAR, R. M., & ALDRICH-BLAKE, F. P. G. 1972. Survival, mating and rearing strategies in the evolution of primate societies. *Folia primat.* 17: 1–19.

HALL, K. R. L. 1962a. Numerical data, maintenance activities, and locomotion of the wild chacma baboon, *Papio ursinus. Proc. Zool. Soc. Lond.* 139: 181–220.

—— 1962b. Sexual, agonistic and derived behaviour patterns of the wild chacma baboons, *Papio ursinus. Proc. Zool. Soc. Lond.* 139: 283–326.

—— 1965. Behaviour and ecology of the wild patas monkey, *Erythrocebus patas. J. Zool.* 148: 15–87.

HAMILTON, W. D. 1963. The evolution of altruistic behaviour. *Amer. Nat.* 97: 354–356.

—— 1964. The genetical evolution of social behaviour. *J. Theoretical Biol.* 7: 1–52.

—— 1970. Selfish and spiteful behaviour in an evolutionary model. *Nature Lond.* 228: 1218–1220.

—— 1971. Selection of selfish and altruistic behaviour in extreme models. In: J. P. Eisenberg and W. S. Dillon (eds.), *Man and Beast: Comparative Social Behavior.* Washington, D.C.: Smithsonian Press.

—— 1972. Altruism and related phenomena, mainly in social insects. *Annual Review of Ecology and Systematics* 3: 193–232.

IMANISHI, K. 1957. Social behaviour in Japanese monkeys, *Macaca fuscata. Psychologia* **1**: 47–54.
—— 1960. Social organization of sub-human primates in their natural habitat. *Current Anthrop.* **1**: 393–407.
ITANI, J., & SUZUKI, A. 1967. The social unit of chimpanzees. *Primates* **8**: 355–381.
ITANI, J., TOKUNDA, K., FURUYA, Y., KANO, K., & SHIN, Y. 1963. The social construction of natural troops of Japanese monkeys in Takasakiyama. *Primates* **4** (3): 1–42.
JAY, P. 1962. Aspects of maternal behaviour among langurs. *Ann. N.Y. Acad. Sci.* **102**: 468–476.
—— 1963a. Indian langur monkey (*Presbytis entellus*). In: C. H. Southwick (ed.), *Primate Social Behaviour*. Princeton: Van Nostrand.
—— 1963b. Mother–infant relations in langurs. In: H. Rheingold (ed.), *Maternal Behavior in Mammals*. N.Y.: Wiley.
—— 1965. Common langur of north India. In: I. DeVore (ed.), *Primate Behavior*, N.Y.: Holt, Rinehart and Winston.
JAY, P., ed., 1968. *Primates: Studies in Adaptation and Variability*. N.Y.: Holt, Rinehart and Winston.
KAUFMANN, J. H. 1966. Behaviour of infant rhesus monkeys and their mothers in a free-ranging band. *Zoologica* **51**: 17–28.
—— 1967. Social relations of adult males in a free-ranging band of rhesus monkeys. In: S. A. Altmann (ed.), *Social Communication among Primates*. Chicago: U. of Chicago Press.
KAWAMURA, S. 1958. Matriarchal social ranks in the Minoo-B troup: A study of the rank system of Japanese monkeys. *Primates* **1**: 149–156. Also S. A. Altmann (ed.), *Japanese Monkeys*. Atlanta: The Editor, 1965. pp. 105–112.
KOFORD, C. B. 1963. Rank of mothers and sons in bands of rhesus monkeys. *Science* **141**: 356–357.
KOYAMA, N. 1967. On dominance rank and kinship of a wild Japanese monkey troop in Arashiyama. *Primates* **8** (3): 189–216.
—— 1970. Changes in dominance rank and division of a wild Japanese monkey troop in Arashiyama. *Primates* **11** (4): 335–390.
KUMMER, H. 1968. *Social Organisation of Hamadryas Baboons*. Chicago: U. of Chicago Press.
KUMMER, H., & KURT, F. 1963. Social units of a free-living population of hamadryas baboons. *Folia primat.* **1**: 4–19.
LEACH, E. 1970. *Lévi-Strauss*. London: Fontana/Collins.
LÉVI-STRAUSS, C. 1963. *Structural Anthropology*. Translated from the French by Clarke Jacobsen and Brooke Grundfest Schoepf. New York: Basic Books.
—— 1969. *The Elementary Structures of Kinship*. Edited by Rodney Needham. Translated from the French by J. H. Bell, J. R. von Sturmer, and Rodney Needham. Boston: Beacon Press; London: Eyre and Spottiswoode.
LOY, J. 1972. The effects of matrilineal relationships on the behaviour of juvenile rhesus monkeys. *Abstracts of the 71st Annual Meeting of the American Anthropological Association.*
MARLER, P. 1969. *Colobus guereza:* territory and group composition. *Science* **163**: 93–95.
MARSDEN, H. M. 1968. Agonistic behaviour of young rhesus monkeys after changes induced in the social rank of their mothers. *Animal Behaviour* **16**: 38–44.
MAYNARD SMITH, J. 1964. Group selection and kin selection. *Nature Lond.* **201**. 1145–1147.
MAXIM, P. E. & BUETTNER-JANUSCH, J. 1963. Field study of the Kenya baboon. *Amer. J. Phys. Anthrop.* **21**: 165–179.
MILLER, M. H., KLING, A., & DICKS, D. 1973. Familial interactions of male rhesus monkeys in a semi-free-ranging troop. *Amer. J. Phys. Anthrop.* **38**: 605–611.

33

MISSAKIAN, E. A. (n.d.). Genealogical mating activity in free-ranging groups of rhesus monkeys on Cayo Santiago. MS. New York: The Rockefeller University.

NISHIDA, T. 1972. Note on the ecology of the red colobus monkeys (*Colobus badius tephrosceles*) living in the Mahali Mountains. *Primates* 13: 57–64.

OMAR, A., & DEVOS, A. 1971. Annual reproductive cycle of an African monkey (*Cercopithecus mitis kolbi*, Neumann). *Folia primat.* 16: 206–215.

POIRIER, F. E. 1970. Niligri langur (*Presbytis johni*) of South India. In: L. A. Rosenblum (ed.), *Primate Behaviour*. Vol. I. N.Y.: Academic Press.

RANSOM, T. W., & ROWELL, T. E. 1972. Early social development of feral baboons. In: F. E. Poirier (ed.), *Primate Socialization*. N.Y.: Random House.

REYNOLDS, V. 1965. *Budongo: A Forest and its Chimpanzees*. London: Methuen.

—— 1966. Open groups in hominid evolution. *Man* n.s. 3: 209–223.

ROWELL, T. E. 1966. Forest-dwelling baboons in Uganda. *J. Zool.* 149: 344–364.

—— 1972. *Social Behaviour of Monkeys*. Harmondsworth: Penguin.

SADE, D. S. 1965. Some aspects of parent–offspring relations in a group of rhesus monkeys, with a discussion of grooming. *Amer. J. Phys. Anthrop.* 23: 1–17.

—— 1967. Determinants of dominance in a group of free-ranging rhesus monkeys. In: S. A. Altmann (ed.), *Social Communication among Primates*. Chicago: U. of Chicago Press.

—— 1968. Inhibition of son–mother mating among free-ranging rhesus monkeys. *Science and Psychoanalysis* 12: 18–37.

SCHALLER, G. B. 1963. *The Mountain Gorilla*. Chicago: U. of Chicago Press.

STRUHSAKER, T. T. 1967a. Behaviour of vervet monkeys. *Univ. Cal. Pub. Zool.* 82.

—— 1967b. Social structure among vervet monkeys (*Cercopithecus aethiops*). *Behaviour* 29: 83–121.

—— 1967c. Ecology of vervet monkeys (*Cercopithecus aethiops*) in the Masai-Amboseli Reserve, Kenya. *Ecology* 48: 891–904.

—— 1967d. Behaviour of vervet monkeys and other cercopithecines. *Science* 156: 1197–1203.

—— 1967e. Auditory communications among vervet monkeys (*Cercopithecus aethiops*). In: S. A. Altmann (ed.), *Social Communication among Primates*. Chicago: U. Chicago Press.

—— 1969. Correlations of ecology and social organisation among African cercopithecines. *Folia primat.* 11: 80–118.

SUGIYAMA, Y. 1964. Group composition, population density and some sociological observations on hanuman langurs (*Presbytis entellus*). *Primates* 5: 17–37.

—— 1965a. Behaviour development and social structure in two troops of hanuman langurs (*Presbytis entellus*). *Primates* 6: 213–247.

—— 1965b. On the social change of hanuman langurs (*Presbytis entellus*) in their natural habitat. *Primates* 6: 381–418.

—— 1967. Social organization of hanuman langurs. In: S. A. Altmann (ed.), *Social Communication among Primates*. Chicago: U. of Chicago Press.

SUGIYAMA, Y., YOSHIBA, K., & PARTHASANATHY, M. D. 1965. Home range, breeding season, male group and inter-troop relations in hanuman langurs. *Primates.* 6: 73–106.

SUGIYAMA, Y., & PARATHASANATHY, M. D. 1969. A brief account of the social life of hanuman langurs. *Proceedings of the National Institute of Sciences, India.* 35, B: 306–319.

TIGER, L., & FOX, R. 1971. *The Imperial Animal*. New York: Holt, Rinehart and Winston.

TRIVERS, R. L. 1971. The evolution of reciprocal altruism. *Quarterly Review of Biology.* 46: 35–57.

—— 1972. Parental investment and sexual selection. In: B. Campbell (ed.), *Sexual Selection and the Descent of Man 1871–1971*. Chicago: Aldine.

VANDENBERGH, J. 1967. The development of social structure in free-ranging rhesus monkeys. *Behaviour* **29**: 174–194.

WASHBURN, S. L., & DEVORE, I. 1961. Social life of baboons. *Scient. Amer.* **204**: 62–71.

YAMADA, M. 1963. A study of blood-relationship in the natural society of the Japanese macaque. *Primates* **4** (3).

ZUCKERMAN, S. 1932. *The Social Life of Monkeys and Apes*. London: Routledge and Kegan Paul.

Norbert Bischof

Comparative Ethology of Incest Avoidance[1]

I INTRODUCTION

The special position held by man in the animal kingdom is usually defined within the framework of the terms 'culture' and 'nature'.

Lévi-Strauss (1970) gives two criteria for this differentiation: (i) Only culture establishes *rules*, natural behaviour being spontaneous; (ii) Cultural characteristics depend on historical coincidence, while only that which is natural in man is observable *universally*.

'In the light of these criteria,' the author continues, 'we are faced with a series of facts which are not far removed from a scandal: we refer to that complex group of beliefs, customs, conditions and institutions described succinctly as the prohibition of incest, which presents [. . .] and inseparably combines, the two characteristics in which we recognize the conflicting features of two mutually exclusive orders. It constitutes a rule, but a rule which, alone among all the social rules, possesses at the same time a universal character' (Lévi-Strauss 1970: 8). He continues: 'Here therefore is a phenomenon which has the distinctive characteristics both of nature and of its theoretical contradiction, culture', and so 'presents a formidable mystery to sociological thought' (ibid.: 10).

Lévi-Strauss attempts to solve this mystery as follows: 'The prohibition of incest is in origin neither purely cultural nor purely natural, nor is it a composite mixture of elements from both nature and culture. *It is the fundamental step because of which, by which, but above all in which, the transition from nature to culture is accomplished: the prohibition of incest is where nature transcends itself*' (ibid.: 24, italics added).

With this idea Lévi-Strauss is clearly following in Sigmund Freud's tradition (see Freud, 1924). Considering the influence exerted on cultural anthropology by these two authors, it is not surprising that similar trains of thought are nowadays prevalent (e.g. Maisch 1968; Wyss 1968).

At the present time comparative ethologists are interested in making the study of nature available for the comprehension of cultural phenomena. This being so, it is evident that the supposition of a point of transition of nature into culture should awaken their interest. This interest gave rise to an investigation on which the following report is based. It should be mentioned in advance that the results so far obtained run counter to prevailing anthropological, sociological and psycho-analytical theories.

II THEORIES ON THE INCEST TABOO

Classification of the theories

As with the distinction made originally between the terms 'culture' and 'nature', the possible explanations of the incest taboo are usually classified under the headings 'biological' and 'sociological'.

It must be borne in mind, however, that such explanations may answer questions of totally different type. Some authors (Homans and Schneider 1955; Slater 1959; Coult 1963) therefore subdivide further according to the categorical form of the causes given for the incest taboo, using the Aristotelian distinction between *causa materialis, formalis, efficiens,* and *finalis.* Here we can confine ourselves, as also Homans and Schneider (op. cit.) have done, to the two last-named categories. In this way we arrive at a fourfold division, in that we first compare the biological and sociological explanations regarding the *final* cause (the reason, motivation, usefulness) of the incest taboo, and then proceed, again dividing the expositions into biological and sociological, to examine the *efficient* cause, that is, the mechanisms which actually ensure abstinence from whatever is forbidden.

Possible final causes of the incest taboo

1 Biological advantages When motivating the prohibition of incest 'biologically', one generally thinks of the danger connected with the increased probability of homozygosity in incest, namely the manifestation of harmful recessive characters. The main supporters of this argument were Morgan (1877), Maine (1883), and Westermarck (1889). Among modern geneticists there is, for example, Lenz (1962) who takes this view. Empirical evidence of 'incest depression', i.e. deficiency symptoms such as retarded growth, lowered immunity and decreased resistance to disease, under-size, short life-expectancy, and reduced fertility among inbred progeny has not only been repeatedly observed in animal experiments (for survey, see Lindzey 1967), but has also been gained from systematic records on humans (Schull and Neel 1965; Adams and Neel 1967).

2 Sociological advantages Alternatively, comprehension of the incest taboo may be attempted through its value in the ready functioning of social institutions. The palette of these theories is wider, and we must limit our inquiry to a few oustanding examples.

(*a*) There is, first, the older opinion of McLennan (1896), Spencer

(1877; 1896), and Lubbock (1870; 1911), according to which the prohibition of endogamy stems from the practice of marriage by capture: wives are valuable possessions, and perpetual conflict within the group can only then be avoided when ownership is apparent; this can be guaranteed if every man provides himself with a wife from outside the group.

Similarly, it has been postulated that the taboo on sexual promiscuity within the nuclear family was necessary to protect the family from internecine strife caused by mutual jealousy (Freud 1924; Malinowski 1927; 1931; Seligman 1929; 1950).

There are, finally, authors who seriously hold that the incest taboo was invented to save hopeless confusion in kinship terminology (cf. however Fox 1967: 57, *et seq.*).

(*b*) While the three foregoing interpretations see a benefit to the nuclear family itself resulting from the taboo, other theories see in it an advantage for the social units one step higher, that is, for those larger groups which, under the effects of the taboo, have been promoted to providing partners.

The best-known such theory has it that with unbridled incest (to which people would in essence tend) no larger social structures could be built up, as over-reaching cultural achievements could certainly not survive in the atmosphere of selfish particularism created by small nuclear families perpetuating themselves (Tylor 1888; Fortune 1932; White 1948, 1959; Murdock 1949; Mead 1950; Lévi-Strauss 1970; Schelsky 1955).

A similar argument is advanced by Parsons (Parsons 1954, 1964; Parsons and Bales 1955): as it should be in the interests of society that the nuclear family produce *mature* scions, it is therefore required of the individual that he summon enough courage to turn his back on the shelter of his family circle, which would tend to keep him infantile, and stand on his own two feet. From this angle incest avoidance appears as something like an enforced documentation of social maturity.

Possible efficient causes of the incest taboo

1 Biological conditions If we now turn to those factors which concretely hinder incest within a society, the 'biological' theory contends that man has an *instinctive* abhorrence of incestuous mating, and the corresponding taboo is a cultural ritualization of this inherited emotional aversion.

In its most naïve form this hypothesis assumes something like a 'voice of the blood' which sounds a warning when relatives meet. Maisch (1968) connects Hobhouse (1912) and Lowie (1920) with this obviously untenable opinion, without apparently having read the articles quoted.

As a matter of fact these authors agree in principle with a theory which must be taken far more seriously, that of Westermarck (1889) and Ellis (1906), according to which innate sexual repulsion is not felt automatically for blood relatives as such, but rather for persons with whom one has been closely associated in childhood. Some modern authors (e.g. Wolf 1966) endorse this view on the basis of new empirical findings (cf. below, p. 60); currently, however, the theory is generally regarded as repudiated (Maisch 1968: 30 *et seq.*), due to criticism by Freud, Lévi-Strauss, and many others, including incidentally Marx.

2 Sociological conditions While the biological theories postulate an inhibition of incestuous activity primarily 'from within', there are sociological theories which predicate repressions 'from without'—that is, repressions which may be internalized secondarily, but originating through the intervention of social partners, an incestuous inclination thereby initially existing on the part of the individual.

As agent of this repressive activity either the entire society may act or else—to name the most prominent example of this group of theories—the jealousy of the parent of the same sex and the unapproachability of the other-sexed parent in the Œdipus situation after Freud (1924).

Arguments against the biological explanations

1 Against the supposition of biological final causes There exist essentially two arguments at present, disputing the contention that incest between close blood relations damages the congenital fitness of the offspring.

(*a*) It is reasoned that genetic disadvantages resulting from inbreeding are certainly not observed with sufficient frequency to justify such a far-reaching prohibition.

(*b*) On the other hand, it is pointed out that inbreeding in itself cannot produce genetic depression; it merely promotes homozygosity and hence the manifestation of *recessive* characters. This is a disadvantage only when the recessive characters themselves are unfavourable, which of course does not necessarily follow.

To be sure, the proportion of unfavourable to favourable characters for recessive genes is indeed higher than for dominant. This significant circumstance seems to be unknown to some authors (e.g. Maisch 1968). The disparity is caused by selection acting constantly upon the dominant genes, whereas in the recessive pool, sheltered by the dominant alleles, all sorts of litter can collect unpenalized. Accordingly, it would indeed make sense to proscribe marriage practices by means of which the sediment of recessive factors is churned up—provided that inbreeding had been formerly suppressed for a considerable length of time, and consequently a biological depreciation of recessive gene material had already occurred. Even so, an inbreeding depression would be a temporary phenomenon only, as natural selection would soon cleanse the—now manifest—recessive gene pool (cf. East 1927). And if, finally, as Lévi-Strauss (1970: 15) assumes, mankind has developed from an ancestry regularly practising incest, there would indeed have been no eugenic reason suddenly to forbid this.

2 Against the supposition of biological efficient causes   The chief arguments against the assumption of an instinctive abhorrence of incestuous mating can be arranged in four groups, of which the first two have indirect, the last two direct, empirical reference.

(*a*) According to Lévi-Strauss (1970: 16) the 'alleged horror of incest can only be manifested when a kinship relationship is supposedly known, or later established, between the guilty parties, and this sufficiently substantiates that its source cannot be instinctive'. And Freud (1924: 149) could not believe, of a biological instinct, that 'it could err so widely in its psychological expression, that instead of blood relatives harmful to reproduction, it took aim at house-mates and fire-side companions who in this respect are perfectly harmless' (author's translation).

(*b*) Freud (ibid.: 149 *et seq*.) voices a second objection, quoting Frazer (1910: 97):

> It is not easy to see why any deep human instinct should need to be reinforced by law. There is no law commanding men to eat and drink or forbidding them to put their hands in the fire [. . .] The law only forbids men to do what their instincts incline them to do; what nature itself prohibits and punishes, it would be superfluous for the law to prohibit and punish [. . .] Instead of assuming, therefore, from the legal prohibition of incest that there is a natural aversion to incest, we ought rather to assume that there is a natural instinct in favour of it.

(*c*) Immediately following this quotation, Freud states rather presumptuously that 'psychoanalytical experience makes the assumption of an inborn abhorrence of incestuous relationship perfectly impossible. It has on the contrary taught us that the earliest sexual impulses of the human child are regularly of an incestuous nature'. Lévi-Strauss (1970: 17) refers to this passage as follows: 'Psychoanalysis, namely, finds a universal phenomenon not in the repugnance towards incestuous relationships, but on the contrary in the pursuit of such relationships.'

(*d*) It is Lévi-Strauss again (1970: 18) who offers a last empirical argument for the cultural foundation of the incest taboos, in calling incestuous mating 'a natural phenomenon found commonly among animals'. Similarly, Wyss (1968: 136) writes 'that the incest taboo [. . .] is agreed by most investigators to be the cultural step which differentiates man from the anthropoids'.

An answer to the first two objections will be given further on (pp. 59 and 62). The nature of the third argument makes analysis extremely difficult, and it will be attempted elsewhere. Thus the fourth argument remains to be tackled now; if correct, this would indeed be of considerable weight.

III SOCIAL STRUCTURES AND INCEST BARRIERS IN MAMMALS

Individual bonding and the necessity for incest barriers

It is characteristic of the fourth argument that its many proponents have hardly ever seriously tested it. Had they done so, they would surely have come up against the empirical fact that *in the whole animal world with very few exceptions no species is known in which under natural conditions inbreeding occurs to any considerable degree.*

This statement is trivial as long as we are dealing with animals having no attachment to conspecifics, or at best only collective–anonymous attachment, and which furthermore are not sedentary. In this case the general diffusion occurring soon after birth makes for ample intermingling. In animals of such a low level of socialization no instinctive incest barriers have been observed: brother and sister cannot single each other out among other conspecifics, and so accept each other readily as sexual partner if they happen to meet.

It is quite different with bonding-motivated animals, however, that is, animals having the ability to recognize each other *individually*, and the *inclination to affiliate with* acquainted conspecifics. This selective preference must generally hit family members, and one could expect that the maturing young would practise sexual activity inside this ready-formed zone of sympathy. *This, however, is precisely what nature systematically avoids*, and the measures adopted will be presented below.[2]

'Bonding motivation' is one of the concrete specifications, necessary for scientific clarity, of the hazy term 'love'. It is in no way synonymous with sexual eroticism, and is probably not even derived therefrom; this emphasis is necessary, as psychoanalysis shows little inclination for such differentiation either in theory or in terminology. The distinctive nature of the bonding motivation has been stressed repeatedly by ethologists (e.g. Fischer 1965; Lorenz 1965) and by ethologically-oriented psycho-analysts (Bowlby 1969). But human psychological research has also reached this conclusion, chiefly in connection with the motivational content analysis of projective techniques ('need for affiliation', see Atkinson 1958).

The following considerations are confined to those animal species evidencing bonding behaviour, at least in the form of attachment of offspring to parent, which corresponds then regularly with parental care of the young. In the space available we must limit the survey substantially to mammals.

Even in such a reduced field, however, an exhaustive report cannot be made. Fairly reliable field observations are available for only a small sample of the species concerned, and what relevant information we can extract for our purpose is nearly always a by-product, as the incest question proper is scarcely ever attacked by fieldworkers. Indeed, it has been rather neglected in ethological literature. Although the issue was raised by Heinroth in 1910, and re-stated by Lorenz in 1943, there are, save for a paper on animal psychology by Brückner (1933), only two more recent dissertations by an ethologist or with ethological cooperation (Kortmulder 1968; Aberle *et al.* 1963, respectively), dealing with the general problem of incest barriers in animals.

A thorough examination of field data so far collected concerning the social life of mammals, and an evaluation of this material especially from the aspect of incest avoidance was first carried out by Bischof and Schottenloher (in preparation). The following is an outline of this work.

A brief survey of mammalian social structures

1 Plain solitary structure In some species the need for individual attachment is confined to infancy; in adolescence, it wanes or is at least reduced to a point at which the partners lose each other. Adults are indifferent or even intolerant to conspecifics, except in the differently motivated periods of mating and, in females, brood-care.

Infants of both sexes leave their parents once they are able to do so. In some species, this process *may* be aided by the mother becoming intolerant towards yearlings when she gives birth to her next litter. Males also, when approaching the mother in a new mating period, may show aggression towards the yearlings and drive them away. This, however, is not as general a phenomenon as was sometimes assumed, and can often be shown to be a surplus mechanism which becomes effective mainly when spontaneous separation does not occur for some special reason.

Figure 1 symbolizes the typical social pattern of a species of this kind. Examples can be found among marsupials (opossum: Reynolds 1952; kangaroo: Caughley 1964) and rodents (e.g. hamster, lemming, squirrel: Eisenberg 1966: Eibl-Eibesfeldt 1951, 1953, 1958). A solitary structure in both sexes is also claimed to be typical of some lower primates such as lemurs (Petter 1965), and of most cats, the latter finding being controversial (cf., e.g., for the tiger: Schaller 1967 and Estes 1969: 68 *et seq.*). In this kind of social organization a particular incest barrier is obviously not required. The waning of the need for attachment in adolescence leads to sufficient separation so that the probability of consanguineous mating is lowered to random level. In what follows this process is referred to as 'isolation'.

2 Male solitary structure Isolation would reduce the probability of inbreeding even if it occurred in males only. As long as all male adolescents leave the family to become solitary the females might as well stay with the mother.

Indeed, there are some species exemplifying this kind of social structure, as demonstrated in *Figure 2*. Of these the coati has been most

43

Figure 1   First non-conjugal structure: Solitary life in both sexes

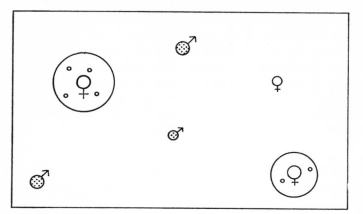

Explanation of symbols (for Figures 1–7):

   ☉♂ ☉♂ = adult and juvenile male

   ♀ ♀ = adult and juvenile female

   ₒ   = infant

Circle = social group, connected by bonding forces (tendency to joint locomotion)
Broken circle (in Figure 6) = privileged subgroup

exhaustively investigated (Kaufmann 1962). The European wild boar
(Gundlach 1968) can also be reckoned in this group.

   Animals of these species can be encountered simultaneously in two
different states of socialization, viz., solitary and in bands. All solitary
specimens are adult males, whereas the bands consist of several females,

Figure 2   Second non-conjugal structure: Multi-female family groups and solitary
           males

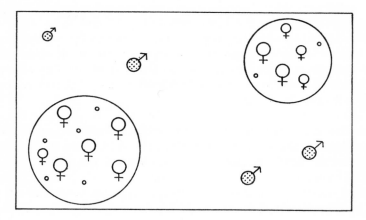

which are presumably all near relatives, and their infants. The young males leave the band before they become sexually mature. In the coati there is evidence that they do so on their own, not being driven away by their mother or by other adults.

It may be noted that the asymmetry of the structure depicted in *Figure 2* is not reversible as far as mammals are concerned. We never encounter the case of females becoming independent and solitary, leaving an all-male family group behind. This seems to be an indirect consequence of the fact that in mammals it is indispensably the mother who nurses the infants.

3 Male cohorts In the coati and, more pronounced, in the wildebeest (Estes 1969) we encounter, as a transient phase, a noteworthy phenomenon: Juvenile and subadult males, on their way to independence, show

Figure 3   Third non-conjugal structure: Multi-female family groups and multi-male cohorts

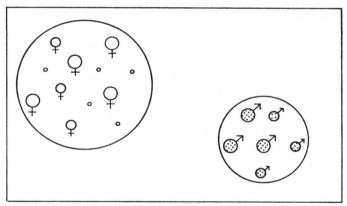

an initial tendency to associate in same-sexed groups. Quite often as, for instance, in most cervids (cf. for the red deer: Darling 1951, Etkin 1964; for the wapiti: Altmann 1963), this tendency persists throughout life. Here, the adult males, as well as the females, form unisexual groups (*Figure 3*). Juvenile males segregate themselves increasingly from their group of origin, but at the same time seek association with others of like sex, so that typical all-male groups are formed, which we call *cohorts* after Chance (1967). Unlike the familiar and quite firmly integrated female groups, they are mostly loosely organized, less intimate, and of variable composition. Such cohorts break up each year during the rutting season, and their members associate with female groups for the duration of sexual activity. During this time the males are intolerant of other cohort members but they nevertheless seek male companionship again when the sexual season is over. In a variant of this structure (e.g. in elephants, cf. Nicholson 1955; Ewer 1968; Hendrichs 1971), the males maintain contact with their cohort even during the rutting season; sex-linked intolerance being replaced by a well-established ranking-order in this case.

It is obvious that with such an organization incestuous mating becomes as unlikely as in the aforementioned social patterns. The mechanism, however, is different now in that the need for social attachment seems to persist up to adult age in both sexes, whereas in males it is obviously the *object* of this attachment which undergoes a *change* before or during adolescence.

For the moment, it must remain a moot point which properties of the old and new attachment objects are decisive in this change from family to cohort. The new companions of an adolescent male are (i) of the same sex and (ii) unfamiliar in so far as they usually do not belong to the group of early childhood association. Either factor could be crucial for the change of object. There might develop a proclivity towards males and an aversion against females; in this case, however, one must additionally postulate an increased readiness to affiliate with unfamiliar conspecifics as well. On the other hand, this second factor alone would suffice to account for the formation of male cohorts, as long as the shift of interest from familiar to strange conspecifics occurs in males only. The absence of females in cohorts would then just follow from a lack of female interest in making new acquaintances.

4 Polygynous structure   The social patterns thus far described can be subsumed under the general label *non-conjugal*. It is characteristic of them that adult males and females do not associate for other than sexual purposes, and then only during a limited period of the year. There are, however, species in which adult animals of both sexes form a lifelong conjugal state. Such a state can occur in the form of polygyny, monogamy, or polygamy.

Polygyny, or harem-formation (cf. *Figure 4*) can be observed in equids like the zebra (Klingel 1967), also in some primates, as, e.g., the hamadryas baboon (Kummer 1957; 1968a, b; 1971) and in a qualified sense, the patas monkey (Hall 1968, Grzimek 1969) and the hanuman langur (Jay 1963; Sugiyama 1967; Yoshiba 1968; Vogel *et al.* 1969). This kind of mating can be formally derived from the non-conjugal structure depicted in *Figure 3* with which it has some features in common, as, for instance, the spontaneous grouping of juvenile, subadult, or (unmarried) adult males in so-called 'bachelor cohorts'. Under the influence of the sexual drive, individuals leave those cohorts and affiliate with females, but once they have done so they stay in a lasting conjugal attachment with them and do not return to cohort life; although a loose contact to former cohort companions may be maintained, as in hamadryas baboons.

Figure 4   First conjugal structure: Polygyny

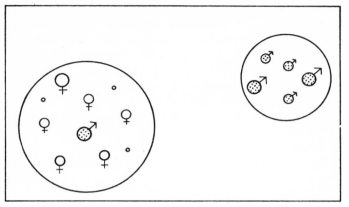

With regard to our question this type of organization poses two problems. First we have to ask the nature of the process which binds the males to the females. Sexual drive would hardly suffice as an explanation. Provided even that the females of the harem come into heat at different times, thus ensuring a certain amount of overlap, there is sufficient indication that the males remain attached also to those females which are momentarily not of interest as sexual partners.

It looks more as if the object of social attachment is exchanged again when a harem is formed. After previously having left their family to associate with other individuals of the same sex, males of these species leave their newly acquired companions for the sake of female conspecifics, which are initially alien, but become after a while as familiar as the members of the original family. We are dealing here with the mechanism of a *double* change of objects. It considerably reduces the probability of mating within the family of origin, which suggests that this is precisely the biological reason for this complicated procedure. Why otherwise should a species which is programmed for lasting matrimony have adolescent males attach themselves in an interim phase to same-sexed peers, instead of just remaining affiliated with the females of their original family?

A second problem that arises as soon as the father stays permanently within the family unit pertains to the destiny of the growing-up daughters. Emancipation of the young males alone would no longer be sufficient to avoid incest now.

Here, indeed, mechanisms have evolved which serve to separate the adolescent daughters from their father. Interestingly enough, in polygynous groups this separation does not seem to occur by means of active emancipation on the side of the female; the activity is rather taken over by young males from outside who abduct the daughters in one way or another.

Female zebras, for instance, when coming into their first oestrus, exhibit a very conspicuous posture. This attracts the attention of young males in the vicinity, who chase these females and cut them off from their family, against the active resistance of the father. One of the males eventually gains the prize, who will usually change her partner several times until she ceases to show her soliciting signals when in heat. Only then does she become uninteresting for cohort males.

In hamadryas baboons the daughters are abducted while still in childhood. Subadult males 'kidnap' girls from neighbouring families, at a stage when the latter are still too young to be of sexual interest to their fathers. The successful 'thieves' have to wait quite a time for the first oestrus of their brides to occur. So, in an initial phase, they just act as caretakers and foster-mothers for them.

Another variant of peaceful taking-over of females occurs also in hamadryas baboons. A harem-leader can sometimes be found accompanied by a so-called 'follower'—a young male whom he has adopted as a kind of apprentice, and to whom he eventually cedes his females.

The same occurs in a much more violent fashion in hanuman langurs, according to Sugiyama (1967) and Yoshiba (1968). At times, a cohort of males invades an existing harem group, and expels or even kills the former leader. Eventually, one member of the cohort even expels his former companions and takes the harem for himself.

5 Monogamous structure It may be asked why, as a rule, in a polygynous group there is place for only one sexually active male, but for several females, although the harem—unlike the all-female family groups of *Figures 2* and *3*—may be composed of individuals who have been initially alien to each other. This asymmetry seems to be caused by a sex-specific intolerance in males which, however, is still lacking in the cohorting phase. Apparently it is bound to a higher maturation level or, as actually in hamadryas baboons, to the presence of females.

If we assume a sex-specific avoidance tendency in females as well, this should result in monogamy. This social pattern occurs occasionally in mammals as, for instance, in the dikdik antelope (Hendrichs & Hendrichs 1971) and, among apes, in gibbons (Carpenter 1940, cf. *Figure 5*). We are not yet able to determine whether this formally very simple, but functionally highly complicated, social structure is due solely to sex-specific aggression. However, as far as the *detachment* of adolescents is concerned, it is indeed reported that this is mostly enforced by aggressive behaviour of the same-sexed parent.

At first glance it may seem that the young remained virtually passive during such a process; that they, for their part, cling to the familiar and secure, only to have maturity thrust upon them by the parent's intervention. Closer observation, however, has shown that often enough the juveniles do make their own positive contribution to the brawl.

48

Figure 5   Second conjugal structure: Monogamy
First variant: Monogamy by separation

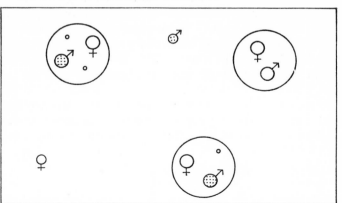

They set the ball rolling by aggression, or at least insubordination, to which the older animals react with increasing impatience.

Here too, apparently, the dissolution of the family seems to be triggered by an 'emancipatory' change in the juveniles; by the building-up of a motivational state, which can perhaps be described as an 'autonomy claim', conceivably analogous to terms such as 'ego-strength' or 'self-confidence' used in human psychology.

It makes sense to assume that the change of object described above is also based upon the growth of this motivational state, that is—to use an anthropomorphic expression—one may ascribe it to increasing 'self-confidence' if the strange and alien is no longer feared but challenged, and if the familiar, which earlier offered security, now engenders merely boredom and surfeit.

It is therefore also in monogamous species that we encounter, behind the superficial spectacle of expulsion, a process of spontaneous object-change, which derives from the double change characteristic of many polygynous species in that the cohorting phase is skipped (see *Table 2*, p. 55).

An *active* component in the process of detachment on the basis of an increasing claim for autonomy is also indicated by the following consideration. In addition to the 'pure' pattern described above, monogamy occurs in a second form. Occasionally, as in wolves (Zimen 1971), marmosets (Rothe, personal communication) and dwarf mongooses (Rasa 1972, and in press) we encounter extended bisexual groups in which, however, only one female and one male are sexually active (see *Figure 6*). This selectivity is warranted by a rank-order that excludes subdominants from propagation.

As yet ethologists have not arrived at full agreement as to the most appropriate definition of rank-order. There is no doubt, however, that

49

this phenomenon must be closely related to the 'claim to autonomy' introduced above. If different individuals form a group, and each of them claims autonomy, i.e. realization of his private interests, then conflicts are bound to ensue. These conflicts can be resolved only if some individuals either give up their group-membership, or change their interests in the direction of a more flexible adaptation to the interests of the others. The latter would mean a reduction of their claim to autonomy.

In some animal societies, compromises of this kind are not equally distributed over all group-members. Instead, a stable asymmetry is built up, in so far as some individuals constantly maintain more auto-

Figure 6 Second conjugal structure: Monogamy
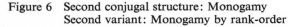
Second variant: Monogamy by rank-order

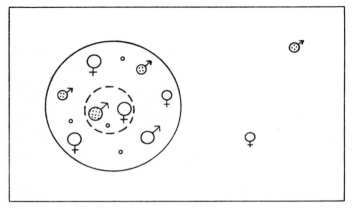

nomy, i.e. more readily exhibit and realize their own interests, as compared with others. It is these cases in which we speak of rank-order. Disagreements concerning the rank position are settled by fighting it out. To make this feasible without disruption of the group, two conditions have to be fulfilled: the aggressor must be appeasable by the symbolic gesture of submission rather than by direct flight on the part of the defeated; correspondingly, the latter should not feel motivated to withdraw, but at least to remain in his place. In fact, he might even show a tendency to attach closely to the aggressor.

This surprising phenomenon has its parallel in the so-called 'identification with the aggressor' which is dealt with in the psychoanalytic theory of neurosis (cf. A. Freud 1936). It may be explained as follows: other things being equal, the consolidation of the claim for autonomy can be taken as a criterion of maturity. Therefore, its revocation when defeated in a rank-order fight can be understood to be a form of regression towards a more infantile stage of social behaviour. According to the above consideration, however, this should also attenuate the readiness to change the object of affiliation. The loser, that is, should return to an attitude of shyness towards strangers and of dependent attachment to the familiar, even though, paradoxically enough, the latter has been the aggressor who initiated the whole process.

Imagine a species which is aggressive in both sexes. Almost inevitably the parents of a growing-up youngster will feel challenged in their superiority when the adolescent displays an ever-increasing autonomy claim by acting in a provocative and inconsiderate manner, or by demonstrating eagerness to change objects. This might indeed incur parental aggression and amount to a generation conflict which has two possible outcomes: either the subadult retracts his autonomy claim and submits himself to parental dominance, which means that a rank-order is established (*Figure 6*), or he maintains his claim until it has become strong enough to bear a change of object. Having reached this point, he emigrates, leaving a monogamous couple of parents behind (*Figure 5*).

This bears substantially on how to interpret the parents' 'chasing away' of the juveniles. Parental aggression alone could scarcely result in family disintegration as long as the young were not yet ripe for it; on the contrary, the effect would more likely be an increase in dependence. If a son's rank-order fight with his father ends with his departure, it shows that for the first time he has *not* knuckled under; the father may have won the fight but he has not managed to curb the son's autonomy claim any longer (cf. also Chance and Jolly 1970: 196 *et seq.*). A strengthened autonomy claim is often indicated, among other things, by sexual activity. Consequently, high-ranking animals tend to interpret courting and copulatory behaviour of subdominant group members as a challenge, and to react by aggressive intervention. This is the reason why in groups structured according to *Figure 6* the rank-order is accompanied by a quasi-monogamous confinement of sexuality to the dominant couple.

The suppression of sexuality in low-ranking group members may occur in two forms. In the simpler case, the need for sexual activity persists in spite of the threat of the group leader, only as a rule the subdominant animals do not *dare* to indulge, but, when they feel themselves unobserved, they may attempt copulation regardless. In another case, the social stress leads additionally to a change of motivation, in that the sexual interest wanes altogether, and may be accompanied by corresponding somatic changes (e.g. reduction of testes, absorption of embryo). This effect, which seems to be present in marmosets (Epple 1966, 1967, 1970; Hampton & Taylor 1970), and tree-shrews (von Holst 1969, 1970) has been referred to as 'psychological castration' (Baldwin 1969). In fact, both mechanisms are effective as incest barriers; the grown-up animal is not able to propagate as long as he stays in his family group. Other than the incest barriers treated above, which are based on spatial separation of potential incest partners before sexual maturity, we encounter here a further group of mechanisms which become effective if the family does not dissolve in time.

6 Polygamous structure This other group of incest barriers, which work by supressing intra-familial sexuality, is of particular importance in the third conjugal structure which is typical of many primate societies (*Figure 7*).

It is characteristic of this kind of social organization that several adult males and females and their progeny are associated in a band.

These bands may be more (rhesus monkeys: Carpenter 1942a, b; Altmann 1962; Koford, 1963, 1965; Kaufmann 1965) or less (chimpan-

Figure 7   Third conjugal structure: Polygamy

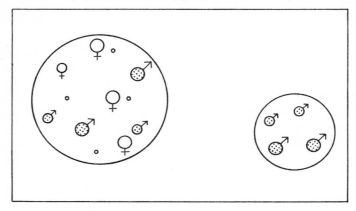

zees: Reynolds & Reynolds 1965; Reynolds 1968; Goodall 1965, 1967; Van Lawick-Goodall 1971: Albrecht & Dunnett 1970) exclusive against strangers. As opposed to the structural type shown in *Figure 6*, the sexual relationships tend to be promiscuous here; they are rather free of jealousy and not confined to high-ranking animals.

A social structure of this kind seems obviously predisposed to promote inbreeding. Nevertheless, there are factors counteracting this. In thorough studies on free-ranging macaques, several investigators observed a striking rarity of mother–son mating (Imanishi 1965; Tokuda 1961–62; Sade 1968). The best-known investigations on this subject have been carried out by Sade (1968) in rhesus monkeys, although, recently, his findings were partly modified by Missakian (n.d.).

Sade considers two mechanisms to be responsible for the reduction of mother–son mating. One of them is a change of object, as described earlier: many males leave their group of origin before becoming sexually mature. The motives for this are not yet quite clear. At any rate, they are not driven away by other males, although in some cases they undergo a loss of rank position shortly before separation. A conflict with the mother is not crucial for the emigration either, because orphans are even more likely to disperse. After leaving their group of origin, some of the young males become solitary, but most of them join bachelor cohorts, until they find a chance to re-enter a mixed-sex band.

Not all young males, however, leave their group before reaching maturity. In those remaining, Sade observed a second incest barrier. Such a juvenile would prefer to be near his mother, with whom he would maintain intimate social contact, entailing close bodily proximity, mutual grooming, and mutual defence. In this association the son displays a permanent *childish* attitude towards the mother, and as long as he does this all *sexual* interaction is ruled out. It looks as if infantile and sexual behaviour were incompatible to a degree, and as if in this case the incest barrier becomes effective by activating the former. We refer to this effect as 'infantilization'.

Missakian (n.d.) has repeated Sade's observations in the same colony but in a different and larger group, and has partly modified them. She also recorded a reduction of mother–son matings, although they occurred somewhat more frequently than in Sade's group. More than two-thirds of mother–son couples proved free of any sexual activity. In the remaining cases the son copulated even exclusively with the mother. Virtually none of those sons, however, had outgrown puberty by more than two years. The older the males grew, the less likely were copulations with their mothers, even if they did not eventually leave them, which, as a rule, they did.

In rhesus monkeys, there is also a striking reduction in the number of observed brother–sister matings. Neither Sade nor Missakian offers an explanation of this phenomenon. Recently though, van Lawick-Goodall (1971) made known her observations on an interesting parallel in chimpanzees. Here—as, incidentally, in some other animal species—sexual play among juveniles is quite common, and in this immature stage familiarity, and therefore relatedness, presents no obstacle (see also pp. 60–1). The author describes, then, how a female who had just reached sexual maturity displayed keen and rather indiscriminate sexual interest in males of casual acquaintance, but at the same time repulsed the advances of her brothers with loud screams, though earlier she had not objected to these advances. We refer to this behaviour as 'repulsion' (when regarded from the standpoint of the male) or 'surfeit-response' (from the standpoint of the female). These terms, to be sure, are only descriptive rather than explanatory. For a first step towards a functional analysis of the underlying processes, cf. Bischof (1972 and in press).

Classification of incest-preventing mechanisms

Table 1 summarizes the incest barriers which have been elaborated in the last section. The principle of classification is twofold: on the one hand, we can discriminate between mechanisms of family dissolution before sexual maturity, and mechanisms which suppress sexuality in those individuals who remain with their family until after puberty; on the other hand, we must separate 'endogenous' from 'exogenous' mechanisms. In the endogenous case, incestuous activity is inhibited due to peculiarities of the individual's own motivational structure. The inhibition is exogenous if the potential incest partner or other individuals build up a barrier, while the individual himself may well be motivated to incestuous activity.

Table 1   Mammalian incest barriers

	Mechanisms of family dissolution	Mechanisms of sexual suppression
exogenous mechanisms	abduction expulsion	threat repulsion
endogenous mechanisms	isolation change of object emancipation	psychological castration infantilization surfeit response

Exogenous mechanisms of family dissolution are the abduction of female adolescents out of polygynous hordes, and the expulsion of adolescents, by aggressive adults. Exogenous suppression of intra-familial sexuality occurs when higher-ranking group members by continuous threat prevent an animal from realizing his sexual intentions, or when the desired partner repulses him. Threat might even generate an *endogenous* inhibition, if it is internalized so as to entail partial or complete extinction of the sexual drive itself (psychological castration). Another endogenous barrier against intra-familial sexuality can be built up by the partner being *too* familiar. This may work either by infantilization, in the sense described by Sade, or by a surfeit reaction, as shown in chimpanzees in Goodall's observations. Finally, endogenous mechanisms of family dissolution have been described under the names of 'isolation' and 'change of object', on which some further remarks will be added below. Also the increasing autonomy claim which leads to the emancipation of adolescents has to be taken into account as an endogenous component in the process of expulsion (cf. p. 51).

The processes occurring in 'isolation' and 'change of object' are listed in *Table 2*. Initially there is always attachment to partners who have become familiar right at the beginning of life due to imprinting-like processes. This normally includes the mother, and sometimes other members of the group of origin as well. At some point during adolescence, a detachment from these individuals of reference will take place, in favour of one of three possible new forms of socialization: the animal

Table 2   Relation between processes of isolation and object-change in males

	infancy	post-infancy	Examples
isolation	FI ——————————→ S ⇌ FS		Hamster
	FI —————→ MS —→ S ⇌ FS		Wildebeest
change of object	FI —————→ MS ←————→ FS		Cervids
	FI —————→ MS —————→ FS		Equids
	FI ——————————→ S —→ FS		Gibbon

Explanations of signs: FI = female imprinting object
MS = male stranger
S = solitary
FS = female stranger
The expression 'stranger' is equivalent to 'not familiarized in early childhood'. It does not exclude later acquaintance. Long-term stages are in bold type.

may associate with novel partners of the same sex; he may become solitary; or he may, as soon as possible, try to affiliate with strangers of the other sex.

These stages are normally passed through in the sequential order indicated by the arrows. As can be seen, some stages may be skipped, naturally with the exception of the last one; at least temporarily, i.e. in the rutting season, affiliation with other-sexed strangers will be attempted. This affiliation may or may not be abandoned afterwards in favour of one of the earlier post-infancy stages. One of the three post-infancy stages will be maintained more or less permanently in adult life, whereas the other two have a transient or episodic character.

Taking into account that the solitary stage is usually characterized by marked territoriality, the question arises whether what we have called 'isolation' is indeed caused by a mere deflation of the need for something to attach to (cf. above, p. 43). Quite possibly we are dealing here with a more complicated process, comparable to a change of object, in which the attachment shifts from the conspecific to the home-range. However, the clarification of this question would require us to unravel the whole problem of the equivalence of partner-bonding and home-range attachment, which must be dealt with elsewhere (cf. also p. 63, note 2).

Missing incest barriers in the animal world

Summing up, it is clear that the die-hard fable of incest tolerance in animals, concocted, according to Maisch (1968: 15), by Diogenes in his tub and henceforth hawked about unscanned, just about turns the empirical findings upside down. There are, to be sure, certain exceptions. 1 First, we must consider that incest occurs habitually in some lower animals with a high reproduction rate, living under ecological conditions which impede or preclude exogamic mating; particularly, that is, in certain *parasites* (mostly mites or worms: Mayr 1963: 408).

2 A further group showing enhanced incest tolerance is formed by the *domesticated animals*; the origination and persistence of the said fable are probably due to these. It is plain that incest barriers are likely to be rudimentary in domesticated animals: the breeder himself will wish to decide which mates with which and when; he has no use for breeding stock which is fastidious. Hence he necessarily exerts a selection pressure tending to breed out possible incest inhibitions.

3 It must be remembered that any interference with the natural living conditions of a species may also disturb instinctive mechanisms and thus reduce their effectiveness. For this reason incest among *zoo animals*, although these are not necessarily domesticated, is less infrequent than in the wild.

Finally, it remains to be noted that a possible selection pressure militating against incest (see below) may sometimes come to a halt at a minimum effect. A mechanism which impeded *all too habitual* inbreeding would suffice; juristic pedantry is not to be expected in nature. The barriers can for example be so low that incest is not made impossible but only improbable; or one of the three possible incestuous combinations (brother–sister, mother–son, father–daughter) could be left open (in polygamous primate groups this is often the last-named); and then again, inhibitions need in principle function with only *one* of the partners, while the other may well incline in vain towards incestuous practice.

IV THE BIOLOGICAL IMPORT OF INCEST AVOIDANCE

On the concept of selectional advantage

The multiplicity of mechanisms restraining incest in the animal kingdom compels us to reconsider the possible existence of a biological final cause, that is, of a selectional advantage in this phenomenon (cf. p. 38 above).

The concept 'selectional advantage' is much too complex to be identified with 'chance of survival'. Such simplifications have fostered the habit of thinking only of hereditary disease in weighing the biological disadvantages attendant upon incest, whereby dismissal of this reasoning is taken as a dispensation from any biological argumentation whatsoever.

In actual fact, a biological value of quite another character can be shown to attach to the incest barriers, i.e. a selection pressure, the power of which dwarfs the small advantage of hereditary fitness in comparison. This selection pressure is identical, as will be explained below, with that which favours *biparental reproduction* above all other forms of propagation.

The selectional advantage of biparental reproduction

To the biological layman the terms 'mating' and 'propagation' seem practically synonymous; nevertheless, propagation can indeed occur without mating throughout the world of organisms including man: that is, it is biologically *possible*.

It appears in three forms (see Hartmann 1956): *asexual reproduction* (agamogenesis), i.e. propagation by division of the whole individual (in protozoa, polyps, and some worms; further in the formation of identical twins) or by budding (found on the very brink of the vertebrate stage); *unisexual reproduction* (parthenogenesis), in which new individuals are produced from unfertilized egg-cells (in some insects); finally *self-fertilization* (autogamy) in hermaphrodites (occasionally observed still in some species of fish).

Thus it is evident that neither fertilization nor indeed propagation necessarily implies the sexual union of two individuals; yet the three above-named forms of monoparental reproduction are remarkably *rare* throughout the vegetable and animal kingdoms. This focuses attention on the biological significance of biparental reproduction: it must have been the outcome of substantial selective forces, as its vulnerability entails so many evident disadvantages.

This biological significance lies, as Weismann realized as long ago as the turn of the century, in the increase of *variety* through the recombination of genetic material.

Evolution is fed by the variability of the species. Only a wide spectrum of distinctive features can ensure, in times of environmental change, that there are enough individuals available who are just then better adapted, and can help the species to pull through the crisis; other, ill-equipped members die out: the species has 'adapted itself to changed conditions' —because it contained sufficient diversity of features, but evidently at the cost of such diversity. Only the constant creation of new variety can save this procedure from grinding to a halt. The source of variety is, after all, the mutation, but this source is a mere trickle. And here, heterogeneous fertilization comes into play, acting as a powerful 'variation-amplifier'.

One can work out what astronomical period of time it would take to effect a somewhat more complex genetic adaptation, if a species were forced to transact all the necessary steps of mutation successively and independently in the *same* germ cell; how much swifter is this process if the 'inventions' are interchangeable between different germ cells! In this sense Mayr (1963: 179) calls recombination 'by far the most important source of genetic variation'.

The selectional advantage of exogamy

The answer to the question of the selectional advantage of exogamy should now be apparent: a species which allowed the obligatory mating of siblings only would retain almost all the disadvantages of biparental generation, without being able to profit from a single one of its advantages. Its variety would sink to the low level of self-fertilization, and its evolutionary rate would accordingly be so halting that it could stand up against competition only under highly favourable conditions of life; as a general rule the lack of adaptive plasticity would act as a death warrant. This means in effect: *existing* species are those which have escaped the danger of obligatory incest, either through favourable circumstances, or through development of special inhibiting mechanisms.

Such mechanisms, however, in the animal species concerned, are integral parts of the genetically fixed instinctive structure, and it would be astonishing if there were not at least rudimentary traces left in man. If so, the biological final cause expounded above would also ultimately be responsible for the universal appearance of the cultural incest taboo.

It must be borne in mind that explaining cultural features as being influenced by natural selection in no way necessarily implies that *cultures* without these features are doomed. We are confronted here with a selectional force which had been operating for untold ages prior to man's emergence, and which had already led to the development of genetically determined motivational structures in the animal kingdom. If any vestiges of these structures still lurk in man's emotional make-up, and he, as with so much that baffles him, has interpreted them mythico-magically, then the cultural taboo emerges indirectly from biological advantages, *without* these last having had a chance to bear fruit in the ridiculously short span of cultural history. In the following section an attempt will be made to establish whether observations on man himself will support this interpretation.

V INCEST BARRIERS IN MAN

Two preliminary questions in an anthropological evaluation

In examining the material basic to Section III we find that some of the incest barriers therein described can be grouped as 'sociological efficient causes', in accordance with the classification introduced on p. 40. Above all, the mechanisms of *abduction* and *threat*, in part also *expulsion*, are 'inhibitions from without'—and of course, from the male point of view, the mechanism of *repulsion*.

Looking at the whole picture, however, these are clearly outnumbered by the 'inhibitions from within' occurring regularly in the species observed, that is, spontaneously developing counter-inbreeding tendencies entered in the instinct inventory of the species. In the higher animals the most important of these are the *change of object, repression of sexuality* and—from the female point of view—*repulsion*; also the mounting of the *autonomy claim* which leads to expulsion.

If we now try to estimate the value of this synopsis for the understanding of man, a twofold question must be asked: first, whether 'inner' inhibiting mechanisms of the kind discussed can be shown to exist in man, too; if so, second, given a background of such mechanisms, how we are to understand the development of corresponding cultural norms. These two issues will be discussed shortly, whereby we shall be able to pick up the threads of the two still unresolved objections to the biological theory of the incest taboo, which were introduced on p. 41.

Emotional avoidance of incest

1 Justification of Westermarck's hypothesis The first of these objections was based upon the naïve surmise that an instinctive aversion to consanguineous mating must be linked, as it were, with a sixth sense for detecting blood relationship: according to this argument, whoever admits the possibility of instinctive incest barriers must necessarily believe in a 'voice of the blood'.

This contention seems incomprehensible, all the more so since Westermarck (1889) and Hobhouse (1912), often quoted ironically in this respect, opposed such conjectures with amazingly modern-sounding arguments.

Contemporary study of instinct does not expect to find nature performing supernaturally. If birds only rarely catch wasps, then the biological reason is that wasps are poisonous. The quality of being poisonous, however, is invisible, and so the mechanism restraining the birds operates, quite simply, as if every insect with black and yellow stripes were a wasp; the hover-fly and other insects with wasp-mimicry have this simplification to thank for their undeservedly carefree lives.

Westermarck therefore advances a legitimate argument, biologically speaking, when he assumes that nature recognizes early-childhood familiarity as a sufficient cue for consanguinity, just as black and yellow stripes stand for poison, the biologically unnecessary inhibition against marrying an adopted sister being tolerated just about as readily as the bird's abstinence from a meal of hover-flies.

2 Endogenous tendencies towards family dissolution In *Table 1* an attempt was made to distinguish between the mechanisms of family dissolution and of suppression of intra-familial sexuality. If we now turn to the first of these, it is easy to find parallels between the psychological alterations of human puberty, on the one hand, and the phenomena of increased *autonomy claim* and *change of object*, as formerly described, on the other.

The more or less radical emancipation of adolescents of both sexes from the child's referential structure of security and obedience—the surfeit with the established order, the lure of the distant, of the exotic, the forbidden, the dangerous—all this is common knowledge in developmental psychology. Even without citing parallel features among animals

59

there can be little doubt that these phenomena are due by and large to *maturation*, although social forces can facilitate, inhibit, or channel them. Still, it may come as somewhat unexpected when, of all things, it is precisely at the bottom of young Oedipus' fight with his father that we find archaic motivational structures whose biological sense is none other than the *prevention* of incest (cf., for a more detailed discussion, Bischof, in preparation).

3 Endogenous suppression of intra-familial sexuality   It is less simple to answer the question whether the phenomena of *inhibition* and *repulsion* of intra-familial sexual activity are also observable in man. Fortunately, however, there is a possibility to test this empirically in societies in which prospective spouses are thrown together as children and grow up together.

Such a culture has been examined by Wolf (1966; 1968) in North Taiwan. Here, among others, two patrilocal marriage forms exist, whose main difference is that in the one the partners come together as adults, whereas in the other the bride is taken into her future husband's family as a child, and the two grow up practically as brother and sister.

The second form of marriage is not esteemed by most young people. This could be partly due to the small prestige actually accruing to this marriage form; yet there are some peculiarities which can scarcely be thus explained, and which have led the author to conclude that such marriages suffer primarily under a disturbance of sexual harmony. At any rate if questions are asked, the repudiation is not ascribed to social disadvantages, but veiled hints are made that such marriages are 'embarrassing' or 'boring'. Adultery of both sexes, concubinage, and recourse to prostitutes are of significantly higher frequency in marriages resulting from child-engagement than in those resulting from adult-engagement.

As moving force behind the arrangement of child marriages Wolf suspects the jealousy of the mother: 'A woman's son is too important in Chinese society for her to accept an intimacy from which she is excluded' (1968: 869). '*The sexual aversion created by the couple's intimate childhood association* [. . .] precludes the development of an exclusive conjugal bond [. . .] The effect [. . .] is to drive a wedge between husband and wife and thereby take the strain off the bonds between the generations' (ibid.: 870, italics added).

A second example is reported by Fox (1962) following Spiro (1958); cf. also Bettelheim (1969) and Shepher (1971). It refers to juvenile development in certain Israeli kibbutzim. The children of a settlement grow up together, grouped separately according to age; living-rooms, dormitories, and bathrooms have, on principle, no separation of the sexes.

Up to about twelve years of age there are no signs of embarrassment between the sexes; on the contrary, the children indulge extensively from an early age in heterosexual play, both in the dormitories and in public. This behaviour is tolerated by the adults in the interests of a repression-free sexual development.

On the threshold of puberty, however, there develops, more markedly in the girls, a mounting tendency to embarrassment, with a considerable admixture of *antagonism* towards the other sex in the same group. The girls reject the co-ed showers and seek to avoid being seen naked by the boys; at the same time their interest turns to young men outside the group.

As far as the authors could discover, no marriages ensued within any one of these peer groups; nor are any cases known of adult sexual relationship of group members. The reason for this abstinence, given by the juveniles themselves, is that they would 'feel like siblings'.

These two instances suggest an obvious parallel to the mechanisms of *inhibition* and *repulsion* of intra-familial sexuality. Other examples point in the same direction, although their substance may not be so apparent at first glance: one illustration is the general damping of sexual activity among the Mountain Arapesh in New Guinea (Mead 1935), who also practise child marriage; another is the report of Rey (1969) according to which celibate professions are preferred by those men having a supernormal attachment to their mothers.

Reverting to the kibbutz example, it remains to be said that here, as with chimpanzees (p. 53) and incidentally as with other mammals and some birds, the incest aversion of puberty is preceded by a period of infantile sexual play with other members of the family. If psycho-analysis, by misapplication of Haeckel's rule that ontogeny recapitulates phylogeny, deduces from this infantile tolerance an 'original' (i.e. precultural) incestuous tendency in man (cf. p. 41) this would be in no wise biologically convincing; for if an incest-aversion should mature rather than be acquired by learning, this process need nevertheless only coincide with the commencement of the reproductive phase, not with the first, still 'harmless' try-out of sexuality in the 'oedipal phase', which may therefore very probably deserve its name.

Nature and culture

1 Cultural ritualization The brief survey undertaken in the last section has already shown that forces are at work in man's motivational make-up which must seriously be taken into consideration as being homologous to instinctive incest barriers. As a rule, however, they appear stylized in the framework of cultural superstructures.

In a comprehensive monograph Cohen (1964: 54 *et seq.*) places the ritualized incest barriers in two groups which are very nearly analogous to the two mechanisms identified above (cf. *Table 1*), viz. (i) family dissolution, and (ii) suppression of intra-familial sexuality.

(*a*) Cohen describes, under the title of *extrusion*, the daily or nightly removal of children aged between eight and ten from their parental homes, and their quartering either with a strange family, in a 'mens. house', a separate hut, or simply in the open. Generally only the boys are extruded, sometimes both sexes, very seldom the girls only

(*b*) The term *brother–sister avoidance* he uses to denote restriction of contact between siblings remaining in the household, as soon as they attain pre-puberty. Siblings may communicate, for example, only through a third person, may not touch or look at each other, or remain together under one roof, etc.

Cohen points out (ibid.: 58–59) that these rites are not merely imposed upon the child, but fall within a stage of development which meets them half-way. Here we see the cultural norm tracing a rather close copy of natural inclinations. From other aspects, however, the original pattern seems to have undergone considerable change. We could expect little success from an attempt to establish a 'biological' explanation for this: cultural anthropology has here its legitimate domain.

Another peculiarity of the cultural incest proscription must be touched upon: its occasional reversal into an incest *pre*scription. The experience of psychoanalysis, it should first be remembered, has revealed that, in coming to terms with emotional tendencies, it seems easier to adopt a contrary attitude than to silence them completely. An explicit command to incest is therefore closer to the universal taboo than is an indifferent tolerance. Moreover, according to Sidler (1971: 9), 'In a monistic world-view, conceiving good and evil as emanating from the same numinous source [. . .] any forcible intrusion upon this numinous sphere, as occurs in the violation of the incest taboo [. . .] can also mobilize healing powers' (cf. also Caillois 1959). One can therefore break a taboo to become taboo, and at least in the case of the incestuous practice still persisting in parts of the Bantu dynasties, it is possible to evidence this motivation (De Heusch 1958).

2 On the function of cultural norms So far, the question raised in the second objection, cited on p. 41, has been left unresolved: why, if natural inhibitions are effective, do cultural ones exist at all?

The answer seems to be that natural inhibitions, as also natural propensities, do not determine but only motivate our behaviour. How we realize them with respect to a given situation, and what compromise we make thereby, has on the whole to be settled by our own initiative, and we are free enough to act contrary to our own nature; but we are not free enough to do so with impunity. We can live at odds with ourselves, and this danger makes us inclined to narrow down the newly gained fullness of scope to within bearable boundaries by means of collectively created norms. Again, however, these norms should keep the emotional field of tension in a sufficiently stable state of equilibrium; and such states cannot be decreed, but must be found.

The creation of cultural norms, therefore, can be regarded as a cognitive achievement, an act of self-interpretation, and these norms will only then remain satisfactory and stable if man is able to recognize his own natural image in this interpretation.

As a rule, to be sure, it will no longer be possible to fathom the original meaning of inherited inhibitions and drives; culture will therefore seek other, more plausible explanations for the emotions which are, after all, there, and demanding their rights, and culture will moreover try to attain other ends by their means. Thus it is quite possible that the various 'sociological' final causes (p. 38 *et seq.*) have all played their part, on a higher level, in the shaping of the incest taboo.

The cogitations of modern structuralists may therefore prove to be an adequate delineation of a superstructure, to lay bare the biological foundations of which has been the object of this report.

Notes

1 This paper is in part based on the following two articles:
The biological foundations of the incest taboo. *Soc. Sci. Inform.* **11** (6): 7–36, 1972.
Inzuchtbarrieren in Säugetiersozietäten. *Homo* **23**: 330–351, 1972.

2 Quite the same problem, incidentally, should arise even in non-social animals, provided they are highly *sedentary*. Such a *substitution* of family attachment by home-range attachment can be observed in some fish and in birds, but there are no examples known in mammals. A *combination* of both forms of attachments, or an ontogenetic *change* is somewhat more frequent. This ought to be mentioned here, although we shall not be able to discuss the implications of home-range attachment and territoriality in the present chapter.

References

ABERLE, D. F., BRONFENBRENNER, U., HESS, E. H., MILLER, D. R., SCHNEIDER, D. M., & SPUHLER, J. N. 1963. The incest taboo and the mating patterns of animals. *American Anthropologist* **65**: 253–265.

ADAMS, M. S., & NEEL, J. V. 1967. Children of incest. *Pediatrics* **40**: 55–62.

ALBRECHT, H., & DUNNETT, S. C. 1971. *Chimpanzees in Western Africa*. Munich: Piper.

ALTMANN, M. 1963. Naturalistic studies of maternal care in moose and elk, pp. 233–253. In: H. L. Rheingold (ed.), *Maternal Behavior in Mammals*. New York: Wiley.

ALTMANN, S. A. 1962. A field study of the sociology of rhesus monkeys *macaca mulatta*. *Annals of the New York Academy of Sciences* **102**: 338–435.

ATKINSON, J. M. (ed.) 1958. *Motives in Fantasy, Action and Society*. Princeton, NJ: Van Nostrand.

BALDWIN, J. D. 1969. The ontogeny of social behavior of squirrel monkeys (*Saimiri sciureus*) in a semi-natural environment. *Folia primatologica* **11**: 35–79.

BETTELHEIM, B. 1969. *Children of the Dream*. New York: Macmillan.

BISCHOF, N. 1972. Inzuchtbarrieren in Säugetiersozietäten. *Homo* **23**: 330–351.

BISCHOF, N. In press. A systems approach towards the functional connections of attachment and fear.

—— In prep. *Das Rätsel Œdipus: Inzesttabu und Generationenkonflikt aus der Sicht der vergleichenden Verhaltensforschung.* Munich: Piper.

BISCHOF, N., & BÖTTGER, H. In prep. Untersuchungen zur Familienauflösung bei Wildgänsen. *Zeitschrift für Tierpsychologie.*

BISCHOF, N., & SCHOTTENLOHER, T. In prep. Group structure and motivational structure in mammalian societies.

BOWLBY, J. 1969. *Attachment.* London: Hogarth Press.

BRÜCKNER, J. H. 1933. Untersuchungen zur Tiersoziologie, insbesondere zur Auflösung der Familie. *Zeitschrift für Psychologie* **128**: 1–110.

CAILLOIS, R. 1959. *L'homme et le sacré.* Glencoe, Ill.: Free Press.

CARPENTER, C. R. 1940. A field study in Siam of the behavior and social relations of the gibbon. Later published in: Carpenter (ed.). Op. cit., pp. 145–271.

—— 1942a. Sexual behavior of free-ranging rhesus monkeys: Specimens, procedures and behavioral characteristics of œstrus. Later published in: Carpenter (ed.). Op. cit., pp. 289–319.

—— 1942b. Sexual behavior of free-ranging rhesus monkeys: Periodicity of œstrus, homosexual, auto-erotic and non-conformist behavior. Later published in: Carpenter (ed.). Op. cit., pp. 319–342.

—— 1965. The howlers of Barro Colorado Islands, pp. 250–292 in: DeVore (ed.). Op. cit.

CARPENTER, C. R. (ed.) 1964. *Naturalistic Behavior of Nonhuman Primates.* University Park, Pa.: Pennsylvania State University Press.

CAUGHLEY, G. 1964. Social organization and daily activity of the red kangaroo and the grey kangaroo. *J. Mammal.* **45**: 429–436.

CHANCE, M. R. A. 1967. Attention structure as the basis of primate rank orders. *Man* **2**: 503–518.

CHANCE, M. R. A., & JOLLY, C. 1970. *Social Groups of Monkeys, Apes, and Men.* New York: Dutton; London: Cape, 1971.

COHEN, Y. 1964. *The Transition from Childhood to Adolescence.* Chicago, Ill.: Aldine.

COULT, A. 1963. Causality and cross-sex prohibitions. *American Anthropologist* **65**: 266–277.

DARLING, F. F. 1951. *A Herd of Red Deer.* London: Oxford University Press.

DE HEUSCH, L. 1958. *Essais sur le symbolisme de l'inceste royal en Afrique.* Brussels: Université Libre de Bruxelles.

DEVORE, I. (ed.) 1965. *Primate Behavior.* New York: Holt, Rinehart and Winston.

EAST, E. M. 1937. Heterosis. *Genetics* **21**: 375 *et seq.*

EIBL-EIBESFELDT, I. 1951. Beobachtungen zur Fortpflanzungsbiologie und Jugendentwicklung des Eichhörnchens. *Zeitschrift für Tierpsychologie* **8**: 370–400.

—— 1953. Zur Ethologie des Hamsters (*Cricetus cricetus L.*). *Z. Tierpsychologie* **10**: 2040–2054.

—— 1958. Das Verhalten der Nagetiere. In: G. H. Helmcke, H. von Lengerken, and D. Starck (eds.), *Handbuch der Zoologie* **10** (13): 1–88. Berlin: Gruyter.

EISENBERG, J. F. 1966. The social organization of mammals, pp. 1–83 in: G. H. Helmcke, H. von Lengerken, D. Starck (eds.), *Handbuch der Zoologie.* Band 8. Berlin: Gruyter.

ELLIS, H. 1906. *Sexual Selection in Man.* Philadelphia, Pa.: F. A. Davis.

EPPLE, G. 1967a. Soziale Kommunikation bei *callithrix jacchus Erxleben 1777*, pp. 247–254 in: D. Stark, R. Schneider, H. J. Kuhl (eds.), *Neue Ergebnisse der Primatologie.* Stuttgart: Fischer.

—— 1967b. Vergleichende Untersuchungen über Sexual- und Sozialverhalten der Krallenaffen (*Hapalidae*). *Folia primatologica* **7**: 37–65.

—— 1970. Maintenance, breeding and development of marmoset monkeys (*callithricidae*) in captivity. *Folia primatologica* **12**: 56–76.

ESTES, R. D. 1969. Territorial behavior in the wildebeest (*Conochaetes taurinus* Burchell 1823). *Z. Tierpsychologie* **26**: 284–370.

ETKIN, W. (ed.) 1964. *Social Behavior and Organization among Vertebrates*. Chicago, Ill.: University of Chicago Press.

EWER, R. F. 1968. *Ethology of Mammals*. London: Logos Press.

FISCHER, H. 1965. Das Triumphgeschrei der Graugans. *Zeitschrift für Tierpsychologie* **22**: 247–304.

FORTUNE, R. 1932. Incest, pp. 620–622 in: E. R. A. Seligman (ed.), *Encyclopedia of the Social Sciences*. Vol. 7. London: Macmillan.

FOX, R. 1962. Sibling incest. *British Journal of Sociology* **13**: 128–150.

—— 1967. *Kinship and Marriage*. Harmondsworth: Penguin.

FRAZER, J. 1910. *Totemism and Exogamy*. Vol. 1–4. London: Macmillan.

FREUD, A. 1936. *Das Ich und die Abwehrmechanismen*. English trans. *The Ego and the Mechanisms of Defence*. London: Imago, 1937.

FREUD, S. 1924. Totem und Tabu. *Gesammelte Schriften*. Vol. X. Leipzig-Zürich: Internationaler Psychoanalytischer Verlag. English trans. *Totem and Taboo*. Standard Edition, Vol. XIII. London: Hogarth.

GOODALL, J. 1965. Chimpanzees of the Gombe Stream Reserve, pp. 53–110 in: DeVore (ed.). Op. cit.

—— 1967. *My Friends the Wild Chimpanzees*. New York: National Geographical Society.

GOODALL, J. VAN LAWICK 1967. Mother-offspring relationship in free-ranging chimpanzees, pp. 287–347 in: D. Morris (ed.), *Primate Ethology*. London: Weidenfeld and Nicolson.

—— 1971. *In the Shadow of Man*. London: Collins.

GUNDLACH, H. 1968. Brutfürsorge, Verhaltensontogenese und Tagesperiodik beim europäischen Wildschwein. *Zeitschrift für Tierpsychologie* **25**: 955–995.

GRZIMEK, B. (ed.) 1969. *Tierleben*. Band 10 (Säugetiere, Teil 1), Band 13 (Säugetiere, Teil 4). Zürich: Kindler.

HALL, K. R. L. 1968. Behavior and ecology of the wild patas monkey in Uganda, pp. 32–120 in: Jay (ed.). Op. cit.

HAMPTON, H., & TAYLOR, A. C. 1970. *Gonadal Development in Marmosets*. (Paper given at the Third International Congress of Primatology, Zürich.)

HARTMANN, M. 1956. *Die Sexualität*. Stuttgart: Fischer.

HEINROTH, O. 1911. Beiträge zur Biologie, namentlich Ethologie und Psychologie der Anatiden, pp. 589–702 in: H. Schalow (ed.), *Verhandlungen des 5. Internationalen Ornithologenkongresses in Berlin, 1910*. Berlin: Deutsche Ornithologische Gesellschaft.

HENDRICHS, H., & HENDRICHS, U. 1971. *Dikdik und Elefanten*. Munich: Piper.

HOBHOUSE, L. 1912. *Morals in Evolution*. London: Chapman and Hall.

HOLST, D. VON 1969. Sozialer Stress bei Tupaias (*Tupaia belangeri*). *Z. vergl. Physiologie* **63**: 1–58.

—— 1972. Renal failure as the cause of death in *Tupaia belangeri* exposed to persistent social stress. *J. comp. Physiol.* **78**: 236–273.

HOMANS, G. C., & SCHNEIDER, D. M. 1955. *Marriage, Authority and Final Causes*. Glencoe, Ill.: Free Press.

IMANISHI, K. 1965. The origin of the human family—a primatological approach. In: K. Imanishi and S. A. Altmann (eds.), *Japanese Monkeys*. Atlanta: privately published.

JAY, P. 1963. The Indian langur monkey (*Presbytis entellus*), pp. 114–123 in: Southwick (ed.). Op. cit.

JAY, P. (ed.) 1968. *Primates*. New York: Holt, Rinehart and Winston.

KAUFMANN, J. H. 1962. Ecology and social behavior of the coati (*Nasua narica* on Barro Colorado Islands , Panama. *University of California Publications in Zoology* **60**: 95–222.

Norbert Bischof

KAUFMANN, J. H. 1965. A three-year study of mating behavior in a free-ranging band of rhesus monkeys. *Ecology* **46**: 500–512.

KLINGEL, H. 1967. Soziale Organisation und Verhalten freilebender Steppenzebras. *Zeitschrift für Tierpsychologie* **24**: 518–624.

KOFORD, C. B. 1963. Rank of mothers and sons in bands of rhesus monkeys. *Science* **141**: 356–357.

—— 1965. Population dynamics of rhesus monkeys on Cayo Santiago, pp. 160–175 in: DeVore (ed.). Op. cit.

KORTMULDER, K. 1968. An ethological theory of the incest taboo and exogamy. *Current Anthropology* **9**: 437–449.

KUMMER, H. 1957. Soziales Verhalten einer Mantelpaviangruppe. *Schweizerische Zeitschrift für Psychologie und ihre Anwendungen* **33**: 1–91. (Beiheft.).

—— 1968a. *Social Organization of Hamadryas Baboons*. Basel: Karger.

—— 1968b. Two variations in the social organization of baboons, pp. 293–312 in: Jay (ed.). Op. cit.

—— 1971. *Primate Societies*. Chicago, Ill.: Aldine.

LENZ, W. 1962. Grundlagen der genetischen Beratung, pp. 3–12 in: F. Linneweh (ed.). *Erbliche Stoffwechselkrankheiten*. Munich: Urban Schwarzenberg.

LÉVI-STRAUSS, C. 1970. *The Elementary Structures of Kinship*. London: Social Science Paperbacks.

LINDZEY, G. 1967. Some remarks concerning incest, the incest taboo and psychoanalytic theory, *American Psychologist* **22**: 1051–1059.

LORENZ, K. 1943. Die angeborenen Formen möglicher Erfahrung. *Zeitschrift für Tierpsychologie* **5**: 235–409.

—— 1965. *Das sogenannnte Böse*. Vienna: Borotha Schoeler. English trans. *On Aggression*. London: Metheun, 1966.

LOWIE, R. H. 1920. *Primitive Society*. New York: Boni and Liveright.

LUBBOCK, J. 1870. *The Origin of Civilization and the Primitive Condition of Man*. London: Longmans.

MCLENNAN, J. F. 1896. *An Inquiry into the Origin of Exogamy*. London. Macmillan.

MAINE, H. J. S. 1883. *Dissertations on Early Law and Custom*. London: John Murray.

MAISCH, H. 1968. *Inzest*. Hamburg: Rowohlt.

MALINOWSKI, B. 1927. *Sex and Repression in Savage Society*. London: Kegan Paul.

—— 1931. Culture, pp. 621–646 in: E. R. A. Seligman (ed.), *Encyclopaedia of the Social Sciences*. Vol. 4. London: Macmillan.

MAYR, E. 1963. *Animal Species and Evolution*. Cambridge, Mass.: Harvard University Press.

MEAD, M. 1935. *Sex and Temperament in Three Savage Societies*. New York: Mentor Books.

MISSAKIAN, E. A. (n.d.) Genealogical mating activity in free-ranging groups of rhesus monkeys (*Macaca mulatta*) on Cayo Santiago. MS. New York: The Rockefeller University.

MORGAN, L. H. 1877. *Systems of Consanguinity and Affinity in the Human Family*. Washington, DC: Smithsonian Institution.

MURDOCK, J. P. 1949. *Social Structure*. New York: Macmillan.

NICHOLSON, B. D. 1955. The African elephant. *African Wild Life* **9**: 31–40.

PARSONS, T. 1954. The incest taboo in relation to social structure and the socialization of the child. *British Journal of Sociology* **5**: 101–117.

PARSONS, T., & BALES, R. F. 1955. *Family, Socialization and Interaction Process*. Glencoe, Ill.: Free Press.

PETTER, J. J. 1965. The lemurs of Madagascar, pp. 292–322 in: I. DeVore (ed.), *Primate Behavior*. New York: Holt, Rinehart & Winston.

RASA, A. 1972. Aspects of social organization in captive dwarf mongooses. *J. Mammal.* **53**: 181–185.

RASA, A. in press. Intra-familial sexual repression in the dwarf mongoose (*Helogale parvula*). *Naturwissenschaften.*

REY, K. G. 1969. *Das Mutterbild des Priesters.* Einsiedeln: Benziger.

REYNOLDS, H. C. 1952. Studies on reproduction in the opossum (*Didelphis virginiana*), *University of California Publications in Zoology* **52**: 223.

REYNOLDS, V. 1968. Kinship and the family in monkeys, apes and man. *Man* **2**: 209–223.

REYNOLDS, V., & REYNOLDS, F. 1965. Chimpanzees of the Budongo forest, pp. 368–425 in: DeVore (ed.). Op. cit.

SADE, D. S. 1968. Inhibition of son–mother mating among free-ranging rhesus monkeys. *Science and Psychoanalysis* **12**: 18–38.

SCHALLER, G. B. 1967. *The Deer and the Tiger.* Chicago, Ill.: University of Chicago Press.

SCHELSKY, H. 1955. Die sozialen Formen der sexuellen Beziehungen, pp. 241–278 in: H. Giese (ed.), *Die Sexualität des Menschen.* Stuttgart: Enke.

SCHULL, W. J., & NEEL, J. V. 1965. *The Effects of Inbreeding on Japanese Children.* New York: Harper and Row.

SELIGMAN, B. Z. 1929. Incest and descent. *Journal of the Royal Anthropological Institute* **54**: 231–272.

—— 1950. The problem of incest and exogamy: A restatement. *American Anthropologist* **52**: 309–316.

SHEPHER, J. 1971. *Self-imposed Incest Avoidance and Exogamy in Second Generation Kibbutz Adults.* New Brunswick, NJ: Rutgers University. (Unpublished doctoral thesis.)

SIDLER, N. 1971. *Zur Universalität des Inzesttabu.* Stuttgart: Enke.

SLATER, M. K. 1959. Ecological factors in the origin of incest, *American Anthropologist* **61**: 1042–1059.

SOUTHWICK, C. H. (ed.) 1963. *Primate Social Behavior.* New York: Van Nostrand.

SPENCER, H. 1877. *Principles of Sociology.* London: Williams and Norgate.

SPIRO, M. E. 1958. *Children of the Kibbutz.* Cambridge, Mass.: Harvard University Press.

SUGIYAMA, Y. 1967. Social organization in langurs, pp. 221–236 in: S. A. Altmann (ed.), *Social Communication among Primates.* Chicago. Ill.: University of Chicago Press.

TOKUDA, K. 1961–62. A study of the sexual behavior in a Japanese monkey troop. *Primates* **3**: 1–40.

TYLOR, E. B. 1888. On a method of investigating the development of institutions: Applied to laws of marriage and descent. *Journal of the Anthropological Institute* **18**: 245–269.

VOGEL, C. 1969. Sozialverhalten indischer Affen. *Naturwissenschaftliche Rundschau* **22**: 383–389.

WESTERMARCK, E. 1889. *The History of Human Marriage.* New York: Allerton Press.

—— 1934. Recent theories of exogamy. *Sociological Review* **26**: 22–44.

WHITE, L. A. 1948. The definition and prohibition of incest. *American Anthropologist* **50**: 416–435.

—— 1959. *The Evolution of Culture.* New York: MacGraw-Hill.

WOLF, A. P. 1966. Childhood association, sexual attraction and the incest taboo: A Chinese case. *American Anthropologist* **68**: 885–988.

—— 1968. Adopt a daughter-in-law, marry a sister: A Chinese solution of the incest problem. *American Anthropologist* **70**: 864–874.

WYSS, E. 1968. *Strukturen der Moral.* Göttingen: Vandenhoeck und Ruprecht.

YOSHIBA, K. 1968. Local and intertroop variability in ecology and social behavior of common Indian langurs, pp. 217–242 in: Jay (ed.). Op. cit.

ZIMEN, E. 1971. *Wölfe und Königspudel.* Munich: Piper.

© Norbert Bischof 1975

N. Blurton Jones

Ethology, Anthropology, and Childhood

INTRODUCTION

The last few years have seen yet another attempted invasion of the human and social sciences by biologists. There is a long but tenuous tradition of biologists moving into the study of human behaviour, with more or less influence on the main stream of the social sciences. Darwin and Galton are widely known as biologists; the biological youth of Freud (who was what nowadays would be called a neuroembryologist) and Piaget (whose freshwater ecology must have been a basis for his ability to do real natural history on human intellectual development) is less widely acknowledged. The contemporary invasion differs radically; quantity is exchanged for quality (applicable in two senses to all but Galton). The contemporary contributions are much more modest than those of these great names but the sheer weight of numbers of people involved must mean that something is afoot. Biologists think they have something to offer, and a number of psychologists and American anthropologists think there may indeed be things that they can usefully take from the methods and theories of biologists. This get-together is well advanced between biologists and psychologists; one can hardly tell from their work which are biologists influenced by psychology, and which are psychologists influenced by biology. But the meeting between biology and anthropology has tended to take the form of a confrontation, in which theoretical issues are indeed at stake but in which confusion and bigotry on both sides are certainly dominant factors. This difference between the meeting of biology—or ethology, to narrow the field somewhat—with psychology, and its meeting with anthropology is partly because different offerings were made by the biologists. Ethologists and psychologists are getting together primarily about methods, and to some extent about theories and data on development of behaviour. Ethologists and anthropologists have met on the more complex issues of man–animal comparisons, and the implications of the evolutionary history of human behaviour.

Ethologists doing direct studies of human behaviour (applying ethological methods) have mostly worked on children (e.g. the papers in Blurton Jones 1972b), though important direct studies on adults have been carried out by Grant (1969, 1972), Eibl-Eibesfeldt (1972),

69

and Kendon (1973). Most of the work on children has been carried out in the UK, but it is beginning to be done in other cultures (e.g. Konner 1972, in press; Blurton Jones and Konner 1973; Freedman and Freedman 1969; Freedman and Keller 1971) and this brings those of us who primarily apply ethological method to the study of human behaviour into contact with anthropologists.

I shall discuss the reasons for ethologists looking at children in other cultures later in this paper. But even in cross-cultural studies interest centres on studying processes and mechanisms of development of behaviour. However, there are also signs that the direct studies and the evolutionary studies are coming together. Freedman's work on differences between newborn babies in different populations is essential reading for anyone attempting to study mother–infant interaction cross-culturally. Konner's work on infancy among the Zhun/twa (!Kung) Bushmen is set in an evolutionary framework, and others are planning direct studies of infancy as part of overall investigations of the relationships between development of behaviour and social organization. It is my feeling that the evolutionary studies are going to play an increasingly important role, though it may not be the role for which they are at present best known.

Perhaps a background sketch of 'ethology' would be helpful before we go further. 'Ethology' in this usage is an abbreviation of 'comparative ethology' which is the English name given by Lorenz and Tinbergen to what in German Lorenz simply called *vergleichende Verhaltensforschung*. The traditional, dictionary use of 'ethology' was probably not implied, though Crook (1970) has argued that this original use is not so far removed from the possible achievements of comparative ethology. Lorenz has argued that behaviour should be studied in the same way as one studies any other feature of an animal. Tinbergen, supporting this view and putting it into practice in a lifetime of field studies and elegant experiments, has pointed out how this implies that biologists must distinguish at least four kinds of 'why' question about behaviour. 'Why does the animal do that?' may mean:

1 What is the motivation, or short-term causation; what made him do it now? The match with physiology is quite clear.

2 What is its development, how did he grow up to be someone who behaves this way? Embryology and the study of learning are the related fields here.

3 What good does it do him? How does it aid survival? Here ethology moves into ecology, population dynamics, and Tinbergen's recent speciality: experimental demonstrations of the survival value of behaviour.

4 How did this behaviour (or its causal mechanisms and developmental mechanisms) evolve? The phylogeny of behaviour, using methods directly modelled on those of comparative anatomy (Tinbergen 1959; Wickler 1967), is here involved.

Added to this rather fundamental feature of ethology are some attitudes to research: behaviour needs to be described before you start trying to explain it, it helps to know something about the animal in its natural environment, an inductive natural history phase can usefully precede deductive experimental studies. Observable behaviour is all that we have for building our theories (together with observed and manipulated environmental change and measures of internal conditions, e.g. endocrine states), and observable behaviour is what we are trying to explain. From this comes an insistence on trying to describe behaviour in ways that other researchers can use, and to quantify behaviour rather than inferred states or motives as the raw data, leading in recent years to a very firm operationalism (see for example Hinde 1954, 1959, 1970). The latter (Hinde 1970) is the major advanced textbook of animal behaviour, and it deals with the core problems in the study of behaviour. The best introductions to the field are Tinbergen (1963) and Manning (1967). Anthropologists will eventually find a great deal of importance to them in a very recent book by Hinde (Hinde 1974).

These characteristics of ethology apply in full measure to the study of social interactions and social organization of animals. They are not confined to the study of individual behaviour. The adaptive significance of various social organizations has been studied intensively by primatologists (e.g. Crook and Gartlan 1966; Crook 1970; Rowell 1972) and for more and more species by mammalologists (e.g. Kruuk 1972), ornithologists (e.g. Orians 1969), and entomologists (Wilson 1971). Rowell's inexpensive and first-rate book is probably the best introduction to this area.

An important result of this symposium could be a dialogue between ethologists and anthropologists and it is important that ethologists be prepared to learn about anthropology as well as that they should be prepared to tell anthropologists about ethology. Ethologists must come to know more about anthropology and anthropologists to know more about ethology in order that both may understand the differences and meeting-points between the two disciplines. But at the outset I should emphasize that at first sight there appears to be a difference in subject-matter; ethologists study the behaviour that you can see people doing, whereas much of what British social anthropologists appear to me to be doing is, much more ambitiously, to study the ways in which people in various cultures organize their thoughts about the world and about themselves. These are very different topics, though not topics that can be held apart for ever. And there are numerous anthropologists whose work could not be defined in this way. For example, Barth (1966) explicitly states that the anthropologist's data are 'a series of regularities in a large body of individual items of behaviour' (and 'our claim must be that we have discovered some non-random frequency distribution in actions').

There is an insoluble problem in setting the limits to the subject-

71

matter of a review of child ethology. What counts as ethology and what does not? This review could include all of (normal) child psychology and an attempt to put these findings into evolutionary perspective, an interest in evolution being one of the most characteristic features of ethology. Yet the distrust which being an ethologist gives one for much of the established 'information' on the behavioural development of children makes this impossible. But conversely, direct ethological studies have made such a small and recent beginning that their results add up to no sort of an account of child development. Consequently I propose to talk about two things. First, I will describe some points of ethological method and illustrate them with recent studies, some done within our own culture but some cross-cultural. It is in the area of methodology that I think many of ethology's potential contributions lie. Then, second, I will have my say on the general issues of evolution of human behaviour and its relevance for those of us whose concern is human behaviour here and now.

Throughout this paper, when I say 'ethologists' I may in fact mean little more than 'me and people whose work I like' though I intend to mean something like 'zoologists who study behaviour and belong to the school that derived from Lorenz and Tinbergen'. Such people do not all agree with everything that Lorenz or Tinbergen ever said. Some of them do agree, however! Ethology being part of science, data have been generated by the outlook and methods initiated by its founders and these data have often led to earlier theories being deposed. There is, in fact, a constant and healthy process of refining methods and un-earthing hidden assumptions in ethology as, one hopes, there is in any science.

I SOME POINTS ABOUT ETHOLOGICAL METHODOLOGY

There is probably no single methodological character that is diagnostic of ethology; an example of any feature that is given as a characteristic of ethology could no doubt be found somewhere in the other behavioural sciences. But the combination of features may be relatively unusual. It is a combination which shares much with anthropology and something with psychology. One could contrast the 'lateral thinking' inductive work of ethologists and anthropologists with the dedication of psychology to the deductive method and its disdain for facts for their own sake. One could also contrast the, some would say excessive, zeal of contemporary ethologists and psychologists for numbers and statistics with the more confident and relaxed attitude of anthropologists.

In this section I pick on and illustrate some features of ethological method in the hope that this will communicate something of the nature of ethology and the habits of ethologists.

Ethologists like to start with direct and simple data

M. J. Konner, an anthropologist with a keen interest in ethology and child development, has worked on Zhun/twa (!Kung) Bushman infancy (1972; in press). Among other things he presents a graph of the proportion of the time that a Bushman child is in bodily contact with its mother at different ages. This is based on records of direct observation of mothers unaware of the approach of the observer (these were 'spot checks' made as he arrived at a family's home to begin a longer, detailed observation of the child by prior agreement with the mother). Plenty of such data is available for non-human primates but when Konner came to seek comparable data from Western cultures he found scarcely any. He even had to cite one study which reported on one single child. It is an extraordinary thing that only for the Bushman is it possible to examine age curves of single items of behaviour like contact with mother, and crying. Nowhere can one see curves for interesting measures like maternal responsiveness (measured in a way independent of the child's demands, as in Hinde's monkey studies (Hinde and Spencer-Booth 1968), which set such an elegant model for human mother–infant interaction studies). This lack of data is in part due to the implicit belief common among psychologists that the data themselves have no meaning or importance, that only if we select from the data on the basis of what we already believe we know can we produce useful results. Thus psychology has carefully precluded advances in knowledge; the data are allowed to speak to us only through a firm censor well versed in the theory of the age.

Ethologists distrust interview data when they are taken to be data about behaviour; they are data about something else

M. P. M. Richards and Judy Bernal, who might prefer to be called social or developmental psychologists, although they both moved into this area from zoology and animal behaviour, are completing a very important study of Cambridge babies and their families. Bernal (1972) reports a number of findings based on prenatal interviews with the mothers and on special diaries kept by the mothers for the ten days following the birth. We all know the difference between the way first and second babies are treated. Ask any mother. Or do we? In the prenatal interviews the mother said they would not spoil the second one, they wouldn't rush to pick it up whenever it cried, they would just let it cry (this *strange* ideal our mothers strive to fit themselves to). But the diaries showed that they responded more quickly than to firstborns, and that the secondborns spend more time out of their cots and with the mother than the firstborns. Unfortunately, Richards and Bernal did not also interview the mothers on this topic post-natally to see whether they still believed they were 'spoiling' the second child less than the first.

73

My impression is that mothers still do not change their ideas after the birth, for the commonest explanation given by British mothers for the greater placidity and 'easiness' of their secondborns is that they had less time to spoil the second one and didn't rush to it at every cry. They seem to be wrong, and one must point out that, in addition, Waldrop and Bell (1966) show that often the later babies can be more placid even at birth.

Ethologists use an unstructured natural history approach, as well as doing experiments

This is a feature shared with both anthropology, where one finds the best natural history in the human sciences, and survey studies in developmental psychology (Davie *et al.* 1972; Douglas *et al.* 1964, 1968). The basic belief here is that the animal, or the data, can generate hypotheses and that these will be better hypotheses than those derived from elsewhere.

One can also argue that it is hard to know whether factors shown to be able to work in experiments actually account for anything in real life. One can argue too that research aimed only at testing hypotheses about human behaviour which are derived from our everyday experiences or cultural knowledge and biases can never fail to prove the obvious. The unexpected associations will be found only by noticing unexpected associations as they occur before one's eyes, or as they are presented by analyses of the data. It may be surprising to field anthropologists to know that there are people who have never had the experience of getting a new idea while gazing at or talking to their subjects and who cannot believe that this happens. The student who says, 'What am I supposed to be finding out?' is a joke in zoology practical classes but is respected in psychology. In the past this may have put very tight limitations on psychology.

Richards and Bernal again are providing some of the best examples of the pay-off of the natural history approach. Their philosophy of collecting many measures at many levels from pharmacological to sociological is highly productive (see Bernal 1972 on social class and milk composition and on feeding interval and success in breast-feeding; Richards and Bernal 1972 on eye movements, drugs in labour, and social class).

Operational definitions of large categories: Is socialization a single process?

Most studies of development of social behaviour simply assume the reality of categories such as 'aggression' or 'attachment' and then proceed to attempt to 'measure' them by ratings on a predetermined scale of the category, or by observing incidents which are then 'coded' into the categories and counted. This procedure has always seemed very alien to ethologists and recently, perhaps partly as a result of contact

74

with ethology, there has been some move away from this in psychology particularly in the US (e.g. Coates 1972; Lewis & Ban 1971; Maccoby & Feldman 1972). A basic point is that if there is such a 'thing' as 'aggression' or 'attachment' then all the possible measures of it should show signs of measuring the same thing; for example, they may vary together. Otherwise there is little point in the category. If two such measures vary independently, investigation of the variation of some measure (or judgement) of both of them lumped together will obscure their relationship to causal variables rather than reveal them. Incredible as it may seem, there have been studies which go so far as to use behaviour of children towards other children as measures of dependency on or attachment to mothers (in many observational situations there is a negative correlation between behaviour to children and to mother).

In addition, it is hard to be sure that a large category, or an unobservable category, or one defined only by examples, is the same in one culture as in another (e.g. 'generosity' which in some cultures is estimated from unsolicited giving and in others from compliance with requests for gifts). It is highly unlikely that such categories will be comparable across cultures when the categories are derived from our everyday feelings and language, and our conscious concerns about behaviour, which are likely to include the most culture-bound things about us. On the other hand, anatomically defined behaviour is going to be directly comparable in a species such as *Homo sapiens* whose basic anatomy varies so little. This is not to say that the motor patterns described by ethologists are present in all races and cultures (although the data of Eibl-Eibesfeldt, 1972, and Ekman, 1972, suggest they are widespread), just that we can easily tell whether they are or not. Differences and similarities between people in different cultures can then very clearly be examined and described in terms of the occurrence and organization of the observable behaviour. Illustrations of the ways this can be done are given in Blurton Jones (1972b), Blurton Jones and Konner (1973), and Tinbergen and Tinbergen (1972).

Basically the same considerations apply to a developmental concept such as socialization. If we start by asking what it is we see that tells us whether socialization is occurring or not (has occurred or not), we free the concept from hidden assumptions (including assumptions found in our culture and not necessarily in others, whether our culture is a scientific discipline, a social class, a tribe, or a nation). We can then examine whether socialization is a unitary process, and whether all the possible measures do appear to be measuring the same things.

Socialization appears to be given a relatively clear meaning by social anthropologists (e.g. Mayer 1970), who are concerned with the way in which the difference between cultures develop (for which the term enculturation is sometimes used), and with the way that a child's behaviour is modified so that it fits into its particular culture. But psychologists use the term 'socialization' even when working within one culture. Their usage therefore seems much more vague and covers a greater diversity of processes. Socialization is used with the implication that social behaviour has to be 'constructed' by the mother (and perhaps occasionally others) out of the 'random' movements of the child (e.g. Maccoby and Masters 1970, discussing maternal influences on the end of the strongest phase of the child's attachment to the mother). This differs from another traditional view of socialization (having much in common with Freudian theory), that the child has to be taught social behaviour and be 'tamed' from an antisocial savage (no such savages ever survived for long, of course). The first view made a good fit with the Thorndike-to-Skinner era of learning theory and with the idea of the *tabula rasa* mind. These views of socialization might seem to have had very culture-bound origins somewhere early in the transition of Anglo-Saxon thought from nineteenth-century authoritarianism towards contemporary *belief* in tolerance of variation and encouragement of individuality. Many uses of the term 'socialization' indicate a supposition that the child must (in several senses!) be moulded into having some social (and sociable) behaviour to save it from being either asocial or destructively antisocial. The emotion versus intellect dichotomy is possibly part of the same pattern of ethnocentric attitudes.

The traditional view of socialization in psychology seems to rest on a false conclusion from the anthropological data and a neglect of the fact that these data revealed tight and complex social organization to be a universal characteristic of all human populations. The fact that social behaviour and child-rearing practices vary from culture to culture implies that variations in social behaviour are associated with variations in child-rearing, but it does not imply that the existence of social behaviour depends on the kinds of features of child-rearing that are seen to vary between cultures. Careful study of 'Western' ethno-developmental psychology might be very helpful in clearing away hidden assumptions. In evolutionary perspective one can argue that much of the meaning of 'socialization' derives from a mismatch between the preadaptations of babies and the 'Western' views on what a mother should do. In fact R. L. Trivers (1974) has taken the evolutionary perspective on socialization a large step further. He has shown that the selective advantage for altruistic behaviour between siblings is greater for the parent (whose genes are equally represented in all its offspring) than it is for either offspring (each of whose genes are less represented in its siblings). There is thus a situation in natural selection that makes for a conflict of interest between the generations, with parents opposing aggression by one offspring towards the other. This makes it all the more clear that one has to test whether socialization pressures are actually effective. Anthropologists assume that they are, parents find them amazingly inefficient, now biologists show that there is the possibility of a selective disadvantage for the offspring in succumbing to *some* of the socialization pressures.

Some research workers keenly involved with ethology have criticized the whole concept of socialization. Bowlby (1969) and Ainsworth (1969) stress the active role of the infant in evoking the interactions and relationships that constitute its social environment. Indeed, there is now a fair amount of evidence from developmental psychology that the infant has a considerable influence on its caretakers. It becomes very important to distinguish 'development of social behaviour' from 'enculturation' (acquisition of behaviour characteristic of a particular culture). Developmental psychologists often use socialization to cover both.

There are probably some very fundamental differences in near-hidden assumptions about development of behaviour between ethologists and anthropologists. It would be useful to get these into the open. How necessary are these assumptions to the work of the two disciplines? Both anthropologists and sociologists do clearly have hidden assumptions about developmental mechanisms and about motivation. It is surely time (a) that these were tested against the findings of ethologists and psychologists and (b) that ethologists and psychologists looked at these assumptions to see if they suggest neglected areas for study, e.g. in the motivation and development of 'front' behaviour.

Modern ethologists and developmental psychologists think of development as having many causes and influences operating at many levels, from genetic biases or even quite specific canalizations, through to verbally transmitted attitudes about behaviour. My impression is that sociologists and perhaps anthropologists concern themselves only with the latter, even to the extent that they feel there is no need to test or explain assumed links between child behaviour and parental attitudes to this behaviour. This contrast often comes up most clearly in discussions on development of sex differences. But also in thinking about changing cultures one should perhaps admit that there are changes at many levels, and not only the attitudinal level, and that transmission of changes can be through a variety of routes. There are many things that influence development besides parental attitudes and parental example.

Individual differences and cultural differences tell us about individual differences and cultural differences

Smith (1972, 1973) looked in two ways at the dimensions of variation of his data on play and social behaviour of nursery school children with results that have profound implications for a basic method of studies of development. He finds (and his findings were replicated by me, Blurton Jones, 1972c) that the dimensions of individual variation of behaviour are quite different from the dimensions of short-term temporal variation. This means that the common practice of using individual differences to tell us about the development of phenomena that show up in temporal variation (like motivational categories such as aggression) can be misleading. We may not be studying the development of aggression—the factors which produce and organize aggressiveness and aggressive behaviour—we will (as Smith and I show) only be studying factors that produce high rates of aggression and also high rates of many other things, including more pleasant social behaviour. If we want to find out the factors influencing *development* of aggressiveness and its structure we must use measures such as individual variation in the factor loadings of component acts of aggression in analyses of temporal variation in behaviour.

Smith's finding is relevant to the discussion of laughter by Douglas (1971) and the use of cross-cultural data for explaining the occurrence of laughter. Douglas discusses the variation from one culture to another in expressiveness, including the amounts of laughter. She concludes that laughter is an integral part of the culture's system of restraint or otherwise over personal expression. But another kind of cross-cultural comparison is possible and leads to an alternative (but not entirely conflicting) view of the causes of laughter. Konner and I are currently analysing data on the association in half-minute observation blocks of a number of facial expressions with a range of other behaviour on London children and Zhun/twa (!Kung) Bushman children. It seems likely that the findings for laughter and its associated behaviour will be the same in both populations and the same as in Blurton Jones (1967, 1972a), van Hooff (1972), and others. Laughter in children is most strongly associated with chasing and fleeing, hitting-at and wrestling. These *associations* probably hold in both cultures despite the differences in frequency of the relevant behaviours.

This association of laughter with rough-and-tumble play suggests that it would be worth pursuing the causes of laughter by further studies like those of Andrew (1963, 1972) and Ambrose (1963). Van Hooff (1972) has discussed the extension of this motivation into adult laughter. These results are about the motivational contexts, short-term causes, and physiology of laughter but they are not about the reasons why people in some cultures laugh more than people in others. Douglas has shown that part of the reason for this is not just that life is funnier in some cultures than in others, but that in some cultures open expression of anything is more or less limited. This implies, in the same way as Smith's findings about individual differences, that we must take this limitation into account in our studies of laughter. But this says nothing about the reasons why people laugh when they do, nor about the reasons for laughing that may be common to all people. Studies of cultural variation tell us about factors associated with cultural variation. They do not *necessarily* tell us about why people do the things all people do. It may be argued that these cross-cultural constants are of no interest to the cultural anthropologist but *someone* ought to be interested in them. They are not obvious ways of behaving. There are many other ways in which man could behave.

'Universals' in behaviour; differences and similarities in the views of ethologists and anthropologists

Ethologists are interested in universals, sometimes because they feel that these are the phenomena in human behaviour that they should be explaining (in terms of evolutionary history and selection pressures, by the use of comparative methods), and sometimes because they feel that to discriminate universals from variables is the way to find out about genetic influences in behaviour development. Most ethologists probably feel that these two quests are interdependent. Some of us do not, and I think it is important for us to keep open minds on many of these issues. This whole area needs exploration not edicts.

My own view at present is that if one found any universals it would mean that the development of these characteristics was resistant to variation in those features of the individual's environment that actually do vary across cultures. It would *not* mean that a culture in which those features varied is inconceivable though it might very well mean that such a culture was unworkable (could not endure), as discussed later. I also feel that for those of us interested in evolution, our first task is to deal with both the constants and the variables only in hunter–gatherer societies and not at first to attempt much with the biologically recent societies (all agricultural, pastoral, fishing, and industrial societies). However, I certainly would not exclude them from biology. The many-pronged threats to the survival of industrial societies, and perhaps even our whole species, make it clear that we are not free of the laws of life.

One major problem for the ethologists who look at child behaviour is that all the observations have been made on British and American children and we know almost nothing of children in other cultures. (This is a problem for ethologists because they are widely held to be interested in 'universal' features of behaviour, and many of them actually are interested in universally human behaviour.) The main exception to this is the excellent data on children of the Zhun/twa (!Kung) Bushmen gathered by Konner (1972, and in press) and Draper (in press). In some respects we now know more about Bushman children than about our own. But this is only one other culture. However, in some ways the available literature on children in other cultures is superior to that on children in our own culture—anthropologists do not seem to be afraid to describe things! But in many respects the information is patchy, unsystematic, and still not descriptive enough to allow for complete comparability from one study to another. Also, and here we reach a major difference in interests and aims already touched on, anthropologists want to know about differences between cultures and to know how to explain them; ethologists tend to want to know about similarities to see what behaviour is widespread across cultures and what is not. I happen to be interested in both and I find it impossible to restrain my interest in what anthropologists do. As someone interested in the development of human behaviour, I find the Whitings's Six Cultures study (1963) is as important to me as it is to any anthropologist (perhaps more), and so are the papers and discussions in the ASA socialization symposium (Mayer 1970). In fact, the Six Cultures study is remarkably similar to what any ethologist like me would do cross-culturally, except that we would like to think that our methods would both shore up these studies and fill in major gaps (e.g. in the mechanism of transmission of maternal attitudes and socialization pressures to the child's behaviour). I, and also probably Richards (e.g. 1971, 1973) and Konner, would argue that unless one measures alike behaviour and attitudes, and economic and physical factors, one cannot hope to demonstrate the relationships between them. Too many studies, having started with strong theoretical preconceptions, have measured one 'independent variable' and concluded that it was the only and ultimate influence on the 'dependent variables' of child behaviour. I think we would all also regard the cross-cultural data as an extremely important source of evidence about developmental processes and mechanisms.

Because ethologists and anthropologists have misunderstood each other on these points in the past, I want to try to explain my position more fully. I am aware that there are anthropologists who seek 'universal' features of human behaviour just as some ethologists do. Indeed, anyone involved in a scientific study of human behaviour must, almost by definition, be in search of some generalizations. But we expect to find our universals at different levels of abstraction. It is unfortunate that we usually expect to find them at only one level. Thus, to take the extremes, ethologists such as myself (1971) and Eibl-Eibesfeldt (1972) look for very small items of motor behaviour in different cultures, expecting to find, first, some very widespread motor patterns and then a similar low-level organization of these patterns, such as the grouping of brow-flash, smile, and gaze aversion that Eibl-Eibesfeldt describes in greetings in many cultures, though we would not necessarily expect the same classes of people to be greeted in all cultures. Ekman *et al.* (1969) provide important information at an intermediate level and Ekman (1972) gives an extremely valuable discussion of the question of universals. At the other extreme we could perhaps put Lévi-Strauss or some of his advocates who seem, to the very uninitiated, to be finding universals almost at the level of the universality of the concept of 'opposite'.

There are many, many possible intermediates between these extremes. Let us take one: that children in all cultures may develop a capacity to quickly adopt, if they choose, the adult behaviour of any culture. Here we have a perfect compromise. Biologists have often spoken of man's use of tradition or acquired social behaviour as one of his most striking features. Yet here is a feature which on its own would permit almost infinite variability in the 'external forms', the resulting social behaviour. I am not trying to sell this particular idea, which is totally unoriginal and begs a whole heap of questions, but I am using it to try to show that there need be no conflict between the viewpoints of a biologist looking for universals and an anthropologist concerned with differences. As I suggested above, the conflict probably stems from our wishes to have exclusive ownership of the one and only key to human behaviour. There can be, and I'd regard it as a safe bet that there are, many universals at quite different levels of complexity. They would still leave ample room for variation, though they might leave one somewhere short of the *tabula-rasa*-plus-social-forces ideal of some sociologists. There is a lot of room for manœuvre between the ideal positions of the traditional camps.

Part of our problem is, of course, that we cannot distinguish the actual variation from the potential variation. Because of this we can easily assume that the startling variation from our own cultures that can be seen in other cultures is an indication of total and infinite variability. But until anthropology began, we had no systematic idea what was possible beyond our own culture. I contend that we still only know what happens, not what doesn't happen, and that perhaps a cross-species perspective would help here. It would help us to attend to the 'obvious' facts, e.g. that there are no ethnic groupings of 'solitary' people ('solitary' animals are not totally asocial but they are overdispersed, unlike human societies, which are characterized by extreme underdispersion), and to many more interesting things, e.g. that relationships between *identified* individuals are very important in man, as they are in some animals, and to lesser extents in others. Wide cross-species studies, not confined to our rather monotypic nearest relatives (*pace* primatologist colleagues), would do more than this, they would tell us the rules that govern societies of creatures that eat, breed, and die. Wilson (1971), working on insects, has set a pattern for this sort of study that badly needs extending to the mammals. Trivers (1971, 1972) has made great headway with a deductive approach, starting from the fact of natural selection, that shows some of the consequences for reproductive and social behaviour of differences in the amount to which each sex puts itself at risk in caring for its offspring. His (and Hamilton's) kind of demonstration of the adaptive value of some feature does not tell us how the feature develops or works, but simply suggests that there must be mechanisms that fulfil this function. The linkages between seemingly different features of behaviour that are uncovered by work such as that of Trivers, and empirical investigations of the existing data (see later in this chapter), imply that there are some combinations of behaviour that cannot and do not (respectively) exist, i.e. that there are some limits to the kind of societies animals can have. A change in one feature, which may itself easily change, will require changes in other features.

One wonders why the search for this sort of data about human societies is not more popular. Why are exercises such as the various Human

Relations Area Files studies and the Six Cultures Study not more popular among anthropologists? If the problem is the standard and comparability of data then a practical solution is obvious. The problem cannot be 'what do we make of the findings' (I've read enough anthropology to see that anthropologists are people with markedly more intellectual fire than certain others in the human sciences) even though this could be a question to ask about 'universals'. One answer to what we do make of universals might, I have to suggest, be found in comparative animal studies; in others words, we are then looking for an explanation of the evolution of this characteristic of man. Thus with our oversimple example of the capacity rapidly to acquire adult behaviour, there is a great deal of psychological work on the factors that influence the development of this capacity in the individual. There is also a long tradition of casual, slapdash, and thoroughly inadequate speculation about how it evolved and under what selection pressures. Almost any zoology or physical anthropology textbook has some of this strewn around its sections on evolution of the brain or the hand, but we ought to be able to do better than that.

II COMPARATIVE STUDIES OF MAMMALIAN CHILD-REARING PRACTICES

In this section I shall be arguing that there are some activities of relevance to the understanding of human societies that biologists should be pursuing more numerously (and numerately) than they are. These are basically an extension of what physical anthropology has always been doing: looking at associations between anatomy and behaviour. Incidentally, this illustrates why a biologist finds it so hard understand the division of anthropology into two separate territories; 'physical' and 'social': in evolution the ecological niche of any animal is a major determinant of its social and other behaviour; selection pressure on its behaviour is a major determinant of its anatomy and physiology. (This is perhaps the place to suggest investigating the construction of biological models for understanding the role of technology in culture change.)

Comparative biological approaches to the study of human behaviour have a long and variegated history. But contemporary exercises in this area have the enormous advantage of being based on a very large body of new data both on animal behaviour and on the palaeontology and archaeology of early man. While major arguments have been put forward, and research, particular in primatology and anthropology, progresses rapidly, most participants in this area would admit that it was still by no means clear what the comparative studies are doing and what their precise implications will be. It seems to me that several different features of the approaches to this subject can be discerned.

Who is compared with whom?

(*a*) Is contemporary man the subject of comparison or is the comparison restricted to non-agricultural man? The *applications* of knowledge gained from comparative studies will be to contemporary industrial and agricultural man. But in many ways comparative studies can more rigorously be applied to pre-agricultural man.

(*b*) Is man compared with one major model in the animal kingdom, e.g. the baboon, as an example of primate adaptations to savannah life, or the chimpanzee, as man's phylogenetically closest relative? This is a debate running currently among anthropologists working in this area. Is man compared with any animal that shows a similarity on the particular feature under discussion? This is a comparison disdained by the more serious workers in the field. Is man compared with all other mammals so as to discern the likely adaptations to hunting animals larger than oneself (social hunting as in wolves, wild dogs, spotted hyenas, and lions) in an omnivorous ape moving into savannah countryside? This is a more strictly zoological comparison, stressing the ecological niche as well as the geographical habitat, and trying to take into account the phylogenetic relationships. It is a more cumbersome approach and one that leads to no immediate answers, but it is the approach that I and many other zoologists would support.

Are we trying to explain or to extrapolate?

(*a*) One may explain human behaviour in terms of its one-time survival value and its evolutionary origins. The information on these is produced by comparative studies carried out by methods based on those of comparative anatomy. Whether or not these explanations also imply things about the development and causation of the behaviour (as Lorenz, 1966, assumes but Bowlby, 1969, does not) is another question.

(*b*) One may wish to use comparative studies to extrapolate to human behaviour as if data on man were missing data. There are two roles for these extrapolations:

(i) Where early man is concerned, it could be useful to keep comparative evidence as an independent line of evidence. The conclusions from this can then be matched with those from archaeology and from studies of contemporary hunter–gatherers. Thus comparative studies could provide another line of argument about the 'representativeness' of contemporary hunter–gatherers.

(ii) Where contemporary agricultural and industrial man is concerned, the extrapolations may be useful (as are straight comparisons that show up differences between man and any other animal) in suggesting major basic patterns to human behaviour or social organization. This role is not unlike that of cross-cultural comparisons and could be of comparable importance. These, like cross-species comparisons, can draw attention to the peculiarity of features we take for granted and to the generalities traceable in all people. They could also suggest ways of perceiving behaviour that step completely outside our cultural pre-conceptions, an aim basic to any scientific approach to behaviour.

What are the implications of comparative studies?

This is where most of the uncertainty and controversy belongs. It is also the source (apart from the dislike of being related to animals, which seems deeply ingrained in our culture, though not in all cultures) of most of the emotion that attaches to this field of study, and there *is* a thick coating of emotion overlying the real uncertainty. Zoologists are in a special position here. Most of them happen to like and respect animals, and to find in the story of evolution the same sort of splendour that some of their opponents presumably see in the theories upon which they were reared. Reactions against comparative studies that refer to man often deny the compassionate motives of those involved in these studies. Proponents of comparative studies wish to reduce war, starvation, and inequality quite as vigorously as their opponents. The dis-agreements are over the nature of the obstacles to these goals.

It is often thought by proponents and opponents alike that the demonstration of an evolutionary history of some feature of human behaviour would carry implications about the mechanism of develop-ment of this behaviour. But, as Bowlby has pointed out, natural selec-tion only requires that an advantageous character develop reliably in the environment in which it is advantageous. The degree of plasticity of development would be subject to quite different selection pressures. The demonstration of the survival value and evolutionary history of behaviour does not imply that its development is unmodifiable. What may be implied is that, since it is the whole animal that must develop and survive, the animal is a web of interrelated adaptations, and the consequences of changing the development of one character may be (but equally may not be) unpredictably far-reaching. Thus variability of development of one piece of behaviour does not imply amorphous-ness, or infinite variability, of the animal's entire behaviour. There could be relatively limited alternative routes and end-points of develop-ment. There must also be wider or narrower limits to variability and greater or lesser canalization of development of each characteristic. But there is little doubt that natural selection can produce almost anything, and if deeply canalized development of some behavioural character was adaptive for long enough in human history it would have occurred. However, none of us here apply the techniques appropriate to establish-ing whether such forms of development occur; indeed, I trust that we never will apply the surgical techniques of neuroembryology to our own species.

A common and controversial aim of comparative studies currently is to use the comparative method to fill in the gaps in the data, in other words to extrapolate from animals to man. Thus comparative study is used as another source of data about early man, to add to those from archaeology and direct studies of surviving hunter–gatherers. However, these extrapolations (concerning both similarities to and differences from other species) can be made in several ways. Hitherto they have usually concerned 'blind' extrapolations on the basis of early man's known ecology and the behaviour of his known relatives. Thus there is debate about just what a savannah-dwelling, partly carnivorous ape would be like: would it resemble more closely a baboon or a chimpanzee? This approach is perfectly adequate, if non-primate data are included, such as those on the other social hunting species (wolf, lion, African wild dog, spotted hyaena; see Kruuk 1972). However, I contend that it may be possible to do better than this. Since in a large part the comparative method consists of a search for correlations between anatomy, physiology, behaviour, and ecology it should be possible to use these correlations to interpret human anatomy and physiology in terms of their functions (including social functions and correlates) to a much greater extent than has been done by traditional human anatomy and physical anthropology. In other words, we have yet to exploit the full potential of the fact that anatomy and physiology are fossilized behaviour. It is important to be clear right away that I am not predicting simple correlations of social behaviour and anatomy, but because both of them reflect the selection pressures acting in any animal's ecological niche there will be relationships between them. But they will be complex. And there are further complexities to anticipate. First, we are dealing with very big systems and there will be many instances in which it will be hard to attribute causal direction to a link (perhaps biologists need to work together with systems analysts in this area, one way in which we need greater numeracy). Second, we also need to take into account the phylogenetic history of each species; earlier phases in its evolution may have excluded certain possible solutions to problems of survival and these exclusions will restrict possibilities at many points through the system. Thus different kinds of animal achieve adequate and surprisingly similar superficial results in quite different ways.

Thus discussions of the behavioural nature of early man that are based on comparisons with other animals so far lack any firm link for the extrapolation. Argument is still at the stage: 'If we have an ape going out into open country he ought to have adapted in a way comparable to a monkey going out to open country, and this is how. . . .' I am suggesting that we try to link extrapolations wherever possible to anatomical and physiological features. So an argument of the above type can be supplemented in this way, taking an example from mammalian child-care (Blurton Jones 1972d): animals who carry their babies have milk with low fat and protein concentrations and, unlike nest-dwelling babies, the offspring are born at the peak of their growth rate.

85

Besides (the old argument) being higher primates, who all, forest or savannah dwellers, carry their babies, humans have milk with low protein and fat concentrations and their babies are born at the peak of their growth rate. Another example might be drawn from the adult end of the life history. It would be interesting to collate data on sex differences in adult size with composition of adult breeding groups. The difference in size of the sexes in man is small compared to many primates, being almost comparable to that in the gibbon (Schultz 1963; Tanner 1962). It seems to me that marked size difference in mammals goes with a high proportion of breeding females to breeding males, but the full data on this need to be assembled and presented (and in some areas the data are poor, e.g. on the group composition of the commonly non-dimorphic S. American monkeys, and in other cases the distribution of the animals in the mating season is complex, e.g. wild sheep, Geist 1971). When sex differences in form, other than size, are also examined for their correlations with behaviour and social organization, one will be able to see what features of social organization go with the size difference in what situations, and then make a *measurably* confident extrapolation of those features of the social organization of early man.

I think that there will be considerable pay-off in information about social organization from pursuing a study of patterns of infant care among the mammals. There are two arguments for examining the mammals as a whole rather than only the primates. (1) Infant-care systems seem to be rather conservative characters and thus give insufficient variation within the primates to allow one to discern the important correlates and thus the functions and roles of the different systems in the social organization. (2) In extrapolating to man as a species one needs data from non-primate groups to enable arguments to be produced about adaptations to a group hunting niche (e.g. differences within the Carnivora between group and solitary hunting species), there being no primates that hunt at a frequency comparable with man.

It is a matter of experience to behavioural ecologists that the adaptations of any one feature of an animal have to be compatible with the adaptations of any other feature. Thus one comes to expect that mammalian 'child-rearing practices' relate to the functioning of the whole social organization. But the relevance of data on infant care to studies of social organization is also made clear in two very important papers, one by Orians (1969) and one by Trivers (1972).

Comparative and evolutionary studies such as those of Eisenberg (1966) and Orians and Trivers have made it quite clear that infant care and characteristics of population replacement and reproduction are major influences on the nature of societies. Thus, Bowlby (1969) has argued that the system of mother and child behaviour that maintains proximity of one to the other has had survival value as a means of protecting children from predators. This is probably true (Konner 1972) and it is apparently highly effective.[1] However, each of the other main systems of mammalian infant care includes adequate safeguards against predation, though some species seem to bear heavier losses of small babies than are found in the carrying species. So there must be reasons why it pays some ungulates to have babies that hide under a bush and get visited by the mother only at long intervals. The long intervals may well be primarily an adaptation against predators, reducing the chance of leading a predator to the baby or leaving scent trails to it (in some species the mother does not go to the hiding-place, the baby comes out to meet the mother when she calls and it returns of its own accord). But why hide the baby in the first place, rather than have it immediately following the mother as in other ungulates? Some traditional answers are fallacious, e.g. that the young are born too soon and too weak to follow. This simply evades the questions. Even if it were the case (it is not, Lent 1973), one would still need to know why it pays to bear the young earlier. Thus questions about survival value lead into a long chain of questions which must gradually elucidate a whole network of selection pressures and compromise solutions to these.

The question of why some animals should have much contact between mother and baby and others little, seems to be poorly answered by some of the more obvious suggestions. One might suppose that this was a way of producing an animal that needed the company of others, or alternatively almost the opposite; that it might produce animals that showed exclusive relationships to only one other individual. A quick look at the mammals shows that there are many social herd-dwelling animals whose babies begin life with minimal contact with the mother. There are animals such as brown bears among which continuous mother–young contact of great duration is seen in species that are markedly solitary as adults. The exclusive relationship theory does not stand up to the test either: there are many species whose young are in continuous contact for a long period that show no such adult phenomenon, e.g. many of those ungulates among which the young follow the mother. The need for mobility of the mother is a more promising possibility, though it might depend on the distances to be covered, with seasonality of movement balanced against duration of infancy. This is an area in which the conservative nature of mammalian child-rearing patterns has to be recalled. For instance, in studying this matter the numerous species of higher primates should receive less attention than the lower primates, which show strikingly varied systems of child-care and thus may more effectively reveal the appropriate selection pressures and correlated behavioural, social, ecological, anatomical, and physiological features. Mobility will in turn depend on a variety of features of the foraging methods of the species and there will be interplay of these with the relationship of child-rearing practices to reproductive rate, parental investment, and mating patterns, which together with predation are major pressures on the demographic and spatial structure of the society.

Examination of the timing of different phases of infancy also looks promising. The total duration of the period of mother–young proximity appears, at a very preliminary glance, to associate both with size of social group and with variety of food sources, as if the mother–young relationship may function with respect either to sociability or to acquiring 'subsistence technology' (foraging techniques). The timing of weaning is not coincident with the end of mother–young proximity, though in some species it seems to coincide with a change in method of transport, and a drop in growth rate (even, apparently, in individuals whose switch to solid food was artifically induced earlier, before the natural weaning age) and sometimes a change in appearance (most obviously in colour patterns). In man, the most conspicuous low point in growth rate seems to coincide with a number of other physiological changes (e.g. reaching a high plateau in haemoglobin levels) and briefly precedes the '5–7 year old shift' described for so many learning and cognitive measures by White (1965). The prediction that around four years old is the time when children walk more than they are carried and will stay home when the mother goes foraging is confirmed for the Zhun/twa Bushmen by Konner (personal communication). However, let us not forget that in London many children stay regularly with people other than the mother before this age, though I would like very much to see data on how much time children spend in their pram or on foot before and after four years of age.

In general, the empirical, statistical approach proposed here would seem to be a useful complement to the largely deductive approach of researchers such as Hamilton (see below) and Trivers (1971, 1972).

SUMMARY

Ethology already meets psychology and can meet anthropology in three general areas: (1) methodology, (2) developmental mechanisms and the question of 'universals' or generalizations, (3) cross-species comparisons and the study of man's evolutionary history.

Ethological methods concentrate on systematic and quantitative study of observable behaviour applied to the causation and the organization of behaviour, to interactions between individuals, and to the organization of groups. Both natural history and experiment are held to be important methods, each with its own role to play. [References to introductory texts are respectively Manning (1967), Hinde and Spencer-Booth (1968) and Hinde (1970), Rowell (1972). Examples of applications to man are in Hutt and Hutt (1970) and Blurton Jones (1972b).]

Sociologists and social anthropologists have implicit theories about the development of behaviour (e.g. that socialization pressures are effective), as well as about motivation. Ethologists have moved away from their early 'preformationist' position to an interactionist view of development, but together with developmental psychologists they have uncovered a great variety of mechanisms of development. Many factors influence development. The ways in which cultures and human behaviour are perpetuated are by no means obvious.

The contribution of comparative studies of man and animals may be rather different from that implied by Ardrey (1961, 1967) and Lorenz (1966). Their potential contribution to the study of human evolution is clear; what is unclear is the value of a knowledge of human evolution for research and knowledge of man here and now—the causation and development of his behaviour and social organization. It is suggested that lack of limits to viable human societies cannot be predicted from variability in the development of single behaviour characteristics. The whole organism has to be considered as a complex system built of adaptations that are compromises between the environment and the adaptations of other features of the animal to the environment.

Note

1 In discussions with Zhun/twa Bushmen about animals and their behaviour, including specific questions about predators on man, Konner and Blurton Jones found the only reports to be of attacks on teenagers and old people, except for people specifically hunting predators, a pattern strikingly similar to the pattern of deaths from predation in baboons. However, actual mortality from predators is *extremely* low in the Zhun/twa.

References

AINSWORTH, M. D. S. 1969. Object relations, dependency and attachment: a theoretical review of the infant–mother relationship. *Child Development*, **40**: 969–1027.

AMBROSE, J. A. 1963. The age of onset of ambivalence in early infancy and indications from the study of laughing. *J. Child Psychol. Psychiat.* **4**: 167–184.

ANDREW, R. J. 1963. The origin and evolution of the calls and facial expressions of primates. *Behaviour* **20**: 1–109.

—— 1972. The information available in mammal displays. In: R. A. Hinde (ed.), *Non-verbal Communication*. London: Cambridge University Press.

ARDREY, R. 1961. *African Genesis*. London: Collins.

—— 1967. *The Territorial Imperative*. London: Collins.

BARTH, F. 1966. *Models of Social Organisation*. (Royal Anthropological Institute, Occasional paper no. 23). London: Royal Anthropological Institute.

BERNAL, J. 1972. Crying during the first ten days of life, and maternal responses. *Devel. Med. & Child Neurology* **14**: 362–372.

BLURTON JONES, N. G. 1967. An ethological study of some aspects of social behaviour of children in nursery school. In: D. Morris (ed.), *Primate Ethology*. London: Weidenfeld and Nicolson.

—— 1971. Criteria for use in describing facial expression of children. *Human Biology* **43**: 365–413.

—— 1972a. Non-verbal communication in children. In: R. A. Hinde (ed.), *Non-verbal communication*. London: Cambridge University Press.

BLURTON JONES, N. G. 1972b. *Ethological Studies of Child Behaviour*. London: Cambridge University Press.

—— 1972c. Categories of child–child interaction. In: N. Blurton Jones (ed.), *Ethological Studies of Child Behaviour*. London: Cambridge University Press.

—— 1972d. Comparative aspects of mother–child contact. In: N. Blurton Jones (ed.), *Ethological Studies of Child Behaviour*. London: Cambridge University Press.

BLURTON JONES, N. G., & KONNER, M. J. 1973. Sex differences in behaviour of London and Bushmen children. In: R. P. Michael & J. H. Crook (eds.). *Comparative Ecology and Behaviour of Primates*. London: Academic Press.

BOWLBY, J. 1969. *Attachment and Loss*, Volume 1. *Attachment*. London: Hogarth Press.

COATES, B., ANDERSON, E. P., & HARTUP, W. W. 1971. Interrelations and stability in the attachment behaviour of human infants. Paper presented at the *Meeting of the Soc. Res. Child Development*. Minneapolis, Minn.

CROOK, J. H. 1970. Social organisation and the environment: aspects of contemporary social ethology. *Anim. Behav.* 18: 197–209.

CROOK, J. H., & GARTLAN, J. S. 1966. Evolution of primate societies. *Nature* 210: 1200–1203.

DAVIE, R., BUTLER, N., & GOLDSTEIN, H. 1972. *From Birth to Seven*. London: Longmans.

DOUGLAS, J. W. B. 1964. *The Home and the School*. London: MacGibbon and Kee.

DOUGLAS, J. W. B., ROSS, J. M., & SIMPSON, H. R. 1968. *All our Future*. London: Davies.

DOUGLAS, M. 1971. Do dogs laugh? A cross-cultural approach to body symbolism. *J. Psychosomatic Res.* 15: 387–390.

DRAPER, P. (in press). The cultural ecology of Bushman childhood. In: R. Lee (ed.), *Studies of Bushman Hunter–Gatherers*, Cambridge, Mass: Harvard University Press.

EIBL-EIBESFELDT, I. 1972. Similarities and differences between cultures in expressive movements. In: R. A. Hinde (ed.), *Non-verbal Communication*. London: Cambridge University Press.

EISENBERG, J. 1966. The social organisation of mammals. *Handbuch der Zoologie*, 8.

EKMAN, P. 1972. Universals and cultural differences in facial expressions of emotion. In: J. Cole (ed.), *Nebraska Symposium on Motivation 1972*. Nebraska: University of Nebraska Press.

EKMAN, P., SORENSON, E. R., & FRIESEN, W. V. 1969. Pan-cultural elements in facial displays of emotion. *Science*, N.Y. 164: 86–88.

FREEDMAN, D., & FREEDMAN, N. 1969. Infant behaviour—differences between Chinese-American and European American new-borns. *Nature* 224: 1227–1969.

FREEDMAN, D. G., & KELLER, B. 1971. Genetic influences on the development of behaviour. In: G. B. A. Stoelinga & J. J. Van der Werff ten Bosch (eds.). *Normal and Abnormal Development of Brain and Behaviour*. Boerhaave Series. Leiden University Press.

GEIST, V. 1971. *Mountain Sheep*. Chicago: University of Chicago Press.

GRANT, E. C. 1969. Human facial expression. *Man* 4: 525–536.

GRANT, E. C. 1972. Non-verbal communication in the mentally ill. In: R. A. Hinde (ed.), *Non-verbal Communication*. London: Cambridge University Press.

HINDE, R. A. 1954. Changes in responsiveness to a constant stimulus. *Brit. J. Anim. Behav.* 2: 41–55.

—— 1959. Unitary drives. *Anim. Behav.* 7: 130–141.

—— 1970. *Animal Behaviour*. 2nd edition. London and New York: McGraw-Hill.

—— 1974. *Biological Bases of Human Social Behavior*. New York: McGraw-Hill.

HINDE, R. A., & SPENCER-BOOTH, Y. 1968. The study of mother–infant interaction in captive group-living rhesus monkeys. *Proc. Roy. Soc. B.* **169**: 177–201.

HUTT, S. J., & HUTT, C. 1970. *Behaviour Studies in Psychiatry.* Oxford: Pergamon.

KENDON, A. 1973. A description of some human greetings. In: R. P. Michael & J. H. Crook (eds.), *Comparative Ecology and Behaviour of Primates.* London and New York: Academic Press.

KONNER, M. J. 1972. Aspects of the developmental ethology of a foraging people. In: N. Blurton Jones (ed.), *Ethological Studies of Child Behaviour.* London: Cambridge University Press.

—— (in press). Maternal care, infant behaviour and development among the Zhun/twa (!Kung) Bushmen. In: R. Lee (ed.), *Studies of Bushman Hunter–Gatherers.* Cambridge, Mass.: Harvard University Press.

KRUUK, H. 1972. *The Spotted Hyaena.* Chicago and London: University of Chicago Press.

LENT, P. C. 1973. Mother–infant relationships in ungulates. In: V. Geist & F. Walther (eds.), *Behaviour of Ungulates and its Relationship to Management.* Morges, Switzerland: I.U.C.N.

LEWIS, M., & BAN, P. 1971. Stability of attachment behaviour and transformational analysis. Paper presented at the *Meeting of Soc. Res. Child Developm.*, Minneapolis, Minn.

LORENZ, K. 1966. *On Aggression.* London: Methuen.

MACCOBY, E. E., & FELDMAN, S. S. 1972. Mother-attachment and stranger-reactions in the third year of life. *Monogr. Soc. Res. Child Development* **37** (1): 1–85.

MACCOBY, E., & MASTERS, J. C. 1970. Attachment and dependency. In: P. H. Mussen (ed.), *Carmichael's Manual of Child Psychology.* New York: Wiley.

MANNING, A. 1967. *An Introduction to Animal Behaviour.* London: Arnold.

MAYER, P. 1970. Introduction. In: P. Mayer (ed.), *Socialization: the approach from social anthropology.* London: Tavistock.

ORIANS, G. 1969. On the evolution of mating systems in birds and mammals. *American Naturalist* **103**: 589–603.

RICHARDS, M. P. M. 1971. Mother–infant interaction and its social context—some comments. In: R. Schaffer (ed.), *The Origins of Human Social Relations.* London: Academic Press.

—— (1973). Social class and the early development of children in Britain. In: *Cultural and Social Influences in Infancy and Early Childhood.* Burg Wortenstein Symposium No. 57. New York: Wenner-Gren Foundation.

RICHARDS, M. P. M., & BERNAL, J. D. 1972. An observational study of mother–infant interaction. In: N. Blurton Jones (ed.), *Ethological Studies of Child Behaviour.* London: Cambridge University Press.

ROWELL, T. 1972. *The Social Behaviour of Monkeys.* Harmondsworth: Penguin Books.

SCHULTZ, A. H. 1963. Age changes, sex differences and variability as factors in the classification of primates. In: S. L. Washburn (ed.), *Classification and Human Evolution.* New York: Wenner-Gren Foundation.

SMITH, P. K. 1973. Temporal clusters and individual differences in the behaviour of preschool children. In: R. P. Michael and J. Crook (eds.), *Comparative Ecology. and Behaviour of Primates.* Londor: Academic Press.

SMITH, P. K., & CONNOLLY, K. J. 1972. Patterns of play and social interaction in preschool children. In: N. Blurton Jones (ed.), *Ethological Studies of Child Behaviour.* London: Cambridge University Press.

TANNER, J. M. 1962. *Growth at Adolescence.* Oxford: Blackwell.

TINBERGEN, N. 1959. Comparative studies of the behaviour of gulls (Laridae): a progress report. *Behaviour* **15**: 1–70.

—— 1963. On aims and methods of ethology. *Z. Tierpsychol.* **20**: 410–433.

TINBERGEN, N., & TINBERGEN, E. A. 1972. Early childhood autism—an ethological

N. *Blurton Jones*

approach. *Advances in Ethology* (supplement to *Z. für Tierpsychologie*) **10**: 1–53.

TRIVERS, R. L. 1971. The evolution of reciprocal altruism. *Quarterly Review of Biology* **46**: 35–57.

—— 1972. Parental investment and sexual selection. In: B. Campbell (ed.), *Sexual Selection and the Descent of Man. 1871.1971*. Aldine, Chicago.

—— 1974. Parent-offspring conflict. *American Zoologist* **14**: 249–264.

VAN HOOFF, J. A. R. A. M. 1972. A comparative approach to the phylogeny of laughter and smiling. In: R. A. Hinde (ed.), *Non-verbal Communication*. London: Cambridge University Press.

WALDROP, M. W., & BELL, R. Q. 1966. Effects of family size and density on newborn characteristics. *Amer. J. Orthopsychiat.* **36**: 544.

WHITE, S. H. 1965. Evidence for a hierarchical arrangement of learning processes. In: L. P. Lipsitt and C. C. Spiker (eds.), *Advances in Child Development and Behaviour*, Vol. 2. New York and London: Academic Press.

WHITING, B. B. (ed.). 1963. *Six Cultures: Studies of Child Rearing*. New York and London: Wiley.

WICKLER, W. 1967. Socio-sexual signals and their intra-specific imitation among primates. In: D. Morris (ed.), *Primate Ethology*. London: Weidenfeld and Nicolson.

WILSON, E. O. 1971. *The Insect Societies*. Cambridge, Mass.: Belknap Press, Harvard.

M. R. A. Chance

Social Cohesion and the Structure of Attention

GROUP COHESION

Since the second world war a great deal of interest has been taken in communication between individuals. This has been approached from electronic theory and through linguistics, and much of the interest in ethology has arisen because it has uncovered the forms of communication in animals and of non-verbal communication in man.

We have to bear in mind, however, that communication is not possible between members of a group unless the group coheres. Individuals have to be within earshot or range of vision, or actually in contact, to communicate at all. This cohesion is based on an infrastructure of relationships which has a primary function of enabling the individuals to stay together, and this is what we must now consider, since it is on this basis that the structure of social behaviour is built.

Every species of sub-human primate has its own unique pattern of social relations which results in a set of subgroupings within the society as a whole, but, as pointed out in *Social Groups of Monkeys, Apes and Men* (Chance & Jolly 1970), three types can be identified in the majority of groups so far studied. These are (1) the assemblies of adult females and their infants, (2) adult cohorts of males, and (3) clusters of juveniles: young who are no longer dependent on their mothers but who have not yet joined either of the adult groups. (Fox 1967; 1972, and Tiger and Fox 1971, Ch. 3, have isolated the same three types.)

These subgroups account for the coherence of their members over large periods of time; but in addition to them there are two distinct modes in which the society as a whole is brought together. These involve the group members paying attention to a focal individual. One of these, which I have called the *hedonic mode*, based on display as a means of gaining attention, is a recent discovery. Chimpanzee and gorilla societies are constructed in this mode. The *agonistic mode*, by contrast, is based on conflict, and on avoidance of attack from the dominant male. Macaque and baboon societies are constructed in this way.

To argue that these two types of society are constructed on these two modes is to suggest that, for the greater part of their lives, individuals are related to each other in these two different ways. The possibility of being persistently aware of a dominant focus arises in the open terrain inhabited by baboons and the rhesus macaque, but is not possible in the forest conditions usually inhabited by the great apes. Monkeys foraging in open country are potentially under attack by predators and need to keep in sight of the dominant males, whereas gorillas moving on the ground through the forest keep in touch by a chain of attention ending up on the focal male. Chimpanzees, on the other hand, moving through the trees, are often separated except by being able to call to one another —and in consequence mutual display is used to reunite the group after its members have come together again.

We shall now go on to consider the features typical of these two modalities as shown by chimpanzees on the one hand and macaques or baboons on the other. Then we shall show that these two systems, by which behaviour is organized, have been found in the behaviour structure of two species of macaque. The great apes are exceptional in that they appear not to be able to cohere by agonistic means.

HEDONIC COHESION

First, let us consider how the cohesion of the whole group is achieved, in an hedonic society, through display. Schaller (1963), in his study of the wild gorilla, was the first to point out that when the dominant male was ready to move off after a rest period, he stood in a prominent position and in a characteristic posture: legs fully extended, vertical and rigid, looking ahead of him. After he had adopted this posture for some time, the rest of the group gathered round him and he then moved off. Schaller also pointed out that when an adult male beats his chest and advances towards an intruder, this not only acts as an intimidating display towards the intruder, but also indicates his whereabouts to the rest of the group.

Reynolds and Reynolds (1965) in their study of the Budongo forest chimpanzees, early on found themselves witnessing the so-called 'carnivals', in which all the chimpanzees hoot, jump about, beat the resonant bases of trees, and wave sticks in the air as they display towards each other in this manner. They noticed that this happened particularly when two groups of chimpanzees met in the forest and that then this mutual displaying was followed by much contact behaviour, including touching, kissing, and some pseudo-sexual behaviour. Reynolds and Luscombe (1969) found that it was much easier to understand the nature of this mutual display when they studied chimpanzees in a thirty-acre enclosure in New Mexico at the time when the American Air Force had a research unit there. On my own visit to this place, I noticed that chimpanzees would wander over the enclosure in the early morning, often separated by 200 yards as they explored in different directions; then perhaps, as a group clambered into the old trees, they would see another group. Immediately, they moved over towards them and made contact, showing some excitement when they met. On one occasion this led to some old sacking and sticks being thrown into the air. Briefly, what Luscombe and Reynolds found was that at times when the fruit was provided early in the morning a lot of mutual displaying took place. At the same time the animals gathered into a group.

They also found that the amount of attention that other individuals paid to any particular chimpanzee depended on his displaying ability, but that the amount of attention did not correlate in any way with the amount of aggressive behaviour shown by any individual. Moreover, an individual's ability to become the focus of attention through displays led him to be the centre of a group of individuals when fruit-sharing took place, and in this way a successful displayer was also successful in getting a large share of the fruit. Subsequently, the group broke up into twos and threes, and these pairs and smaller groups moved off in pursuit of different activities.

As I pointed out earlier, the great apes are peculiar in that they do not show any *persistent* agonism, and this means that, apart from the coherence of small groups, the whole group is brought together only through mutual display binding the individuals' attention from time to time on to one another.

AGONISTIC COHESION

The mechanism that holds agonistic-type societies together is of a different kind. As pointed out earlier, because baboons and macaques live largely in open country, it is possible for them to be aware of the dominant male at all times, and for him thereby to be the focus of the group's attention. Although they may be largely preoccupied by foraging and so be spread out over quite a large area, they are nevertheless ready at any moment to coordinate their own behaviour with that of the dominant male in defence of the group as a whole. As individuals, they do not move out of sight of the centre, so we must assume that they are always attentive to it. This has been demonstrated in the way the hamadryas baboon and the rhesus macaque behave as the troops move away from their sleeping-sites in the morning. Here the direction of the movement of a troop as a whole results from the initiative of sub-adult males moving out in different directions, and from the eventual coordination of this initiative of theirs with the direction the dominant male ultimately decides to take.

Kummer's study (1968) of the one-male groups of the hamadryas baboon, moreover, makes it clear that this unit is the result of the confinement of the individual's attention within his own group, focused on the male. In this way, baboons move about as a very tightly packed group. Here their attention is coordinated by the infliction of neck bites by the male on the members of his harem if they stay any distance away.

From time to time in semi-captive colonies—and during fairly prolonged episodes in the wild—threatening, chasing, fleeing, and various forms of submission are shown in vigorous agonistic encounters between adults of the group, mainly the males. At all times the individual in an agonistic community must be ready to defend his status or avoid being the object of severe attack, and this can only be done by persistent awareness of the dominant focus. The dominant males, having a super-fluity of aggressiveness, may be provoked either accidentally or by transgression of certain spatial and behavioural regularities. Hence an agonistic society is held together by a persistent awareness of the centre, based on a readiness to behave agonistically at any moment. It is not surprising, therefore, that spacing-out and status are persistent pre-occupations of the individuals of such a group. Most interactions within the group (e.g. submission, flight) are designed to terminate the conflict inherent in the situation. In this respect, therefore, agonistic behaviour contrasts markedly with hedonic behaviour, since it tends to terminate active interactions, whereas hedonic behaviour tends to promote social intercourse.

COMPARISON OF HEDONIC AND AGONISTIC BEHAVIOUR

In order to make a comparison between the agonistic and hedonic modes, it is easier to describe one of them and to contrast this description with the corresponding features of the other. In so doing I shall attempt to abstract from the information available the characteristic features of the hedonic mode. I shall do this by contrasting the behaviour of the chimpanzee, which exhibits the hedonic mode in its most characteristic form, with the agonistic behaviour shown by the baboons and macaques.

Undoubtedly, the feature that has the most far-reaching implications is the flexibility of behaviour in the hedonic mode. This leads to an increase both in the range of behaviour and in the readiness with which a switch in behaviour can take place. Thus the chimpanzee extends the variety of behaviour which it exhibits in the wild, and also displays a readiness, in semi-captive situations, to acquire entirely new forms. This may well reflect an enhanced ability to receive rewards for achieving a new skill. This, in turn, undoubtedly reflects a greater flexibility in the control of attention and the individual control of excitement.

One of the outstanding features of chimpanzee society is the prominence of contact behaviour, as illustrated by Vernon Reynolds (1973). In the first place, greetings involve contact behaviour in a number of ways: for example, touching hands, touching another's body, kissing, various forms of hugging, and, frequently, when greatly excited, some form of sexual behaviour. This is often followed by the group's breaking up into pairs of chimpanzees who move off with their arms round each other, leading to the separation of the various subgroups of the society and providing the opportunity for a later reunion.

Since the members of an agonistic society are constantly aware of each other and in each other's presence, the opportunities for meeting in the way the chimpanzees do does not arise. The meeting between two individuals in an agonistic society always involves a heightening of the latent conflict present, and the individual meets this situation by postural and facial appeasement gestures such as presentation, lip-smacking, and 'appeasement face'. If any physical contact is established, it leads to ritual grooming, in which one individual grooms the other; but occasionally mutual grooming takes place among the group. In the forms of contact shown by the chimpanzee, however, there is a great fluidity. Ritual grooming is very stereotyped and consists of sharp downward and sideways combing movement of the fingers of one hand, while the other hand is used for examining and picking at the exposed skin. While pairs of (or even three) subadult chimpanzees will be seen moving over fairly large areas in an exploratory fashion with their arms round each other, the subadult macaques or baboons have no contact, except that the younger subadults may jump on the backs of elder males in periods of social excitement.

From the work of Mason (1965) it is possible to infer that contact in all these instances is a form of reassurance. Undoubtedly, the widespread occurrence of contact in the behaviour of the chimpanzee reduces tension and enables the individual rapidly to control his excitement.

Chimpanzees in the wild are offered a great variety of ways of using sticks as tools. Sticks or branches are used to hit the ground or to throw at potential predators, and also for poking into termite hills to extract the termites, and in these circumstances awkward projections are broken off. Suitable twigs may be found by the discarding of many until a usable one is found or until one is fashioned in the way described.

Macaques and baboons do not use any instruments in the wild. Hence the whole range of problem-solving behaviour which is open to the chimpanzee (studied by Köhler 1927, see Chance 1960, and Schiller 1957, in semi-captivity) is potentially unavailable to these species in the wild. A startling recent discovery by Menzel (1972) has shown that a group of young chimpanzees who had grown up together in semi-captivity were able to develop and jointly exploit a new method of using a pole as a ladder. In all these studies, the ability of the chimpanzee to adjust the use of the instrument to the potentiality of the situation—in order not only to reach a set objective but also to use an invented skill for more than one purpose—indicates clearly that the chimpanzee is capable of extending its repertoire further into the human range than any other subhuman primate. Groups of chimpanzees can considerably expand their repertoire by invention, whereas groups of baboons and macaques are much less likely to.

A notable feature of this expansion of repertoire is the switch of attention from social to non-social aspects of the environment—clearly seen in the mutual inspection of an object by two chimpanzees. Two chimpanzees will sit next to each other and one will watch the other trace the line of a crack in the floor with his finger or a stick, and will momentarily indicate the shared attention between them by running his finger along the crack.

A greater control of attention through an awareness of external features is particularly well illustrated by a mother's awareness of the situation of her young. Both gorilla and chimpanzee mothers are able to anticipate and protect their infants from potentially dangerous situations (see Chance and Jolly 1970). A macaque or baboon mother protects her infant from potential danger only by restricting its activities or responding to its calls.

It seems that, from studies of other species of monkey and baboon, provided they are withdrawn from their social context, they are potentially capable of exhibiting some, if not many, of the problem-solving capabilities of chimpanzees, but only as individuals (Köhler 1927; Chance 1960; Bolwig 1963). Many rudimentary elements of problem-solving and inventiveness can be found in some species with basically agonistic societies; for example, long-tailed macaques (Angst 1971).

The limitations on individual behaviour inherent in the mechanisms designed to reduce conflict rigidify the behaviour of predominantly agonistic species (such as the rhesus macaque and the baboons) to an extent which precludes possibilities based on variability. Agonistic societies contain a very large element of conflict, both in the relationships between individuals and the state of the individuals' motivation. As a result, 'spacing out' is a way of reducing the impact between the two individuals. So also is a 'cut-off' act. This is a distraction, in the simplest instance, of the visual awareness of another individual by deflecting the gaze. Very often, it means paying attention to something else and not simply closing the eyes, for example. This brings about a continual interruption of attention, so preventing ongoing attention to a single feature of the environment, or the mutual inspection of the same object by two individuals so typical of the hedonic mode.

In the agonistic mode, displacement of behaviour may take the form of scratching in the ground, for example, but if so, this is rigid and repetitive, not controlled by an awareness of the environment—as when a chimpanzee moves its fingers along a crack in the ground. The fixity and rigidity of these displacement activities and deflections of attention originate more from the value of these actions in curtailing the conflict between individuals, than from an intrinsic interest in the alternative object towards which attention is directed.

Moreover, in this way attention is frequently deflected to another individual of lower status. This may then be used to displace aggression by threats or actual chasing, as in the following example (Sade 1965).

On April 10th, 1962, for no apparent reason, 1956-Male-I attacked an unrelated adult female while she was drinking. Immediately, the dominant male of the group, Old-Male-A, attacked 1956-Male-I, who fled. Old-Male-A chased him round for about half a minute or more; from a bush, up a palm tree and down the other side, round a building and down a trail. Suddenly, out from under another bush 1956-Male-I's parent, Old-female I, came running on her hind legs carrying her five day old infant. She ran to stand at her son's side, and together they made violent threat gestures and vocalisations at a part of the area empty of all monkeys and observers, and away from the dominant male. Old-Male-A stopped chasing to look at what they seemed to be threatening, then chased 1956-Male-I again, who again threatened loudly away from Old-Male-A. Old-Male-A sat down three feet away peering again in the direction in which 1956-Male-I was threatening. Old-Female-I climbed to sit between them and her son immediately sidled up to her, sat touching her and groomed her. She walked away and 1956–Male-I followed her closely, grooming her whenever she stopped, leaving Old-Male-A peering at nothing. Parents often defend their offspring from the attacks of monkeys more dominant than themselves by diverting the attention of the attacker in just the manner described.

BI-MODAL POTENTIAL IN THE MACAQUE REPERTOIRE

So far, we have described the difference between the hedonic and the agonistic mode in terms of the behaviour of species which are organized predominantly in one or other of these two different modes, and we have abstracted the characteristics of each mode. Following my suggestion (Chance 1967) that rank-ordered behaviour was best understood in terms of the structure of attention, Virgo and Waterhouse (1969) have found, from their study of the Bristol Zoo macaque colony, that there were two structures of attention, each focused on a different individual in the colony. They showed that the structure of attention based on agonistic behaviour was focused on an adult female, and the other, based on grooming and affiliative relations (sitting next, etc.) was focused on an adult male. This immediately drew my attention to earlier observations of my own at Bristol Zoo where I observed that changes of the leadership of the colony followed a period of vigorous display by the future dominant male. Hence it became clear that the features of both modalities are present and employed by members of a single species—the rhesus macaque—and therefore the propensities for these two behavioural features were present in the individual rhesus macaque. Current studies under way with Tom Pitcairn and Walter Angst are providing evidence of the same potentials in the behaviour structure of the long-tailed macaque. We are nearing the point, therefore, when we are able to say of the sub-human primates that an individual possesses a potential for developing these two modes.

What, then, are the essential consequences for the understanding of behaviour structure? Since the hedonic mode is alone capable of enabling behaviour to be constructed entirely on the basis of external information, and the stereotype of the agonistic mode is in such marked contrast to it, we are forced to conclude that the agonistic mode is constructed from fixed action patterns and is therefore programmed from internal sources.

HUMAN COUNTERPARTS

In our discussion of the infrastructure of social cohesion and control we have suggested that, depending on the basic modality in which individuals are operating, the type of communication between them will differ. In human relationships, communication of the type we have been discussing is considered 'non-verbal'. By its nature, this type of communication is concerned with the relationship of the communicator with other individuals in a group, or, as I shall call this attribute of an individual, his primary-group relations.

The infrastructure of social communication is posture, gesture, and facial expression as well as tone of voice, and constitutes the framework of control within which linguistic intercourse takes place.

In an important sense the relationship of individual to individual in a group is tacit, and the way the individual becomes competent in this sphere is through practice in social relations. These are the implications of Bernstein's work, about which I am going to speak, for he has discovered that language can be subordinated to, and used only as an extension of, this tacit relationship.

But Michael Argyle (1972), Adam Kendon (1967), and Paul Ekman (1972) have shown that, in the conversation of literate individuals, eye contact and head and hand gestures are used to control the attention between speaker and listener. This occurs in what we colloquially know as 'talking to one another'. Therefore, language is used in two quite different ways, depending on how it is related to the infrastructure of interpersonal relations.

Both Bernstein (1965) and Jensen (1968), in different ways which are incidental to their main interests, have uncovered the fact that, through differentiation in development (ontogeny), different mental faculties can be subordinated to, and come to be part of, this infrastructure controlling relationships within a group.

Both have used class differentiations as a means of studying the operation of 'intelligence', and Bernstein has been specifically interested in uncovering the fact that differences in intelligence of pupils is related to the patterns of social behaviour found in the family and their general social background. But this is only incidental in Jensen's work.

Bernstein suggests that the limitation of intelligence in some classes of pupil arises from the fact that they possess a *restricted code* of lan-

100

guage, which, through its own structure, indicates that it is used as an extension of the non-verbal social control system which the individual has grown up to consider the essential framework within which language itself is to be used. As an example, children commenting on a strip cartoon of children playing ball and accidentally kicking it through a window-pane, used the pronouns 'you', 'they' and 'we'. 'They' and 'we' indicate out-group and in-group identities and 'you' the speaker's awareness of a single individual within the group.

Nouns describing the objective existence of the components of the cartoon, the ball and the people involved, the window, etc., are not used, and, because of this lack of specification, the language operates only because there is acceptance by the viewer of what is referred to, so that elaboration is redundant. It leaves so much unsaid that it needs the picture to be present in order to make intelligible the meaning of what is said. The picture, in this instance, stands for the shared social situation.

Significantly enough, in this context, the pronoun 'I' is not used, since the person does not see himself as an entity in the total situation. The situation and the action taking place within it are merely an experience *of his*. He does not see himself as part of an objective situation.

An *elaborated code*, on the other hand, renders explicit what is implicit: through the use of nouns which can be modified by adjectives, it increases the flexibility of the verbal repertoire and leads eventually to the development of a fully syntactic language.

Restricted codes are rigid by virtue of their being clamped on to a social situation. There are hints throughout Bernstein's work that individuals who grow to adolescence with a restricted code form social groups in which strong control is the primary concern of members; that is to say, a strong social identity among those controlled, and control of the group by a dominant leader. In personal discussion with Bernstein, he agreed that rank-ordered social structure was typical of those using restricted codes. Elaborated codes—because their use of nouns describes objects and persons in terms of their identity rather than their relationship to the speaker—are based on a form of language which is concerned with the assessment of external reality as opposed to the social reality of experience within a group. An elaborated code is therefore appropriate to behaviour supported by the hedonic mode, and presumably arises out of it. This is consistent with the suggestion in what Bernstein says about the use of the child's imagination as the basis for the construction of an elaborated code. An example of an agonistic human social group would appear to be provided by a social group using restricted codes, since these codes are an extension of the system of social relationship, which is rank-order.

The social relationships of individuals using a restricted code are, therefore, essentially similar to the agonistic control structure of a sub-human primate group. I think, therefore, this demonstrates that such a system can develop into a controlling mechanism in human society.

Evidence that an individual's capacity to learn is also related to whether or not he is operating primarily in the context of a social group comes from the work of the psychologist Jensen. He was able to distinguish between individuals who are primarily associative learners and others who are able to develop powers of classification and abstraction.

Jensen is primarily a psychologist, and therefore only really interested in the fact that these two faculties can appear in different individuals, and in what educational situations these occur. Having demonstrated this distinction, however, he has noticed that powers of simple deduction (associative learning) go together with competence on the playground and concern with social relations, as against the handling of abstract information and classification. Because he is not primarily concerned with the patterns of behaviour that go with these powers of learning, the behavioural distinction is not as clear as we would like it to be, but that there are individuals primarily concerned with social relations, and others concerned with these only as a part of their social awareness, is a distinction which I confirmed through personal correspondence with him.

It looks very much as if here we have the discovery of different types of learning which may be linked to restricted codes, because they are both linked to the development of patterns of behaviour which also restrict the individual to operating within a small or primary group.

The use of language in a restricted code is therefore the subordination of language to an agonistic social structure, in which the information content of the linguistic system is limited, the potential quality of language left undeveloped, and the mental faculties not fully developed. If this argument is correct, then it leaves still to be defined the social structure which permits the full realization of the capability of language to convey objective relationships in the sense usually ascribed to scientific statements. Such a social structure is latent in the hedonic mode, in which social relations are not fixed, and, therefore, are not part of a system specifically functioning to ascribe a role to the individual's behaviour. In such circumstances the individual's attention is free to be guided by interest in external features; an essential requirement for investigating the properties of the external world and the extension of this process into man's pursuit of scientific knowledge.

Tiger and Fox (1971) make a remarkably similar point in their discussion of education. 'The crux of the matter is very simple: dominance and attention are political devices, not instructional ones. Children learn best when their attention is wholly on what they are doing rather than on the teacher. Attention to the teacher and all the neural and emotional processes that this involves may even interfere quite directly with learning' (pp. 166–7). Their insight into the 'anti-intellectual' nature of 'initiatory' systems based on hierarchy and attention to a dominant person (the teacher), and their contrast with an 'instructional' system based on attention to the environment, match my distinction between the agonistic and hedonic modes quite closely.

One thing we should note, therefore, before we go on to consider further examples, is that an agonistically structured group consists of individuals primarily concerned with their social relations, whereas the hedonic structure permits varying degrees of awareness of the physical environment and the intrinsic nature both of other individuals and of the nature of the physical environment.

Moreover, the experience of the physical environment, as well as of other individuals, is, in the agonistic mode, *necessarily* an extension of the social experience. Hence if we see that the primary concern of a person in a group is with maintaining the fixity of the social relations and hence the control of others in that group, and that the individual's awareness of objective reality arises out of his or her social relations and not out of concern for the people themselves, we are probably witnessing the operation of the agonistic mode in these individuals. This seems to me to be the basic characteristic of the Hell's Angels brotherhood.

A report in *The Times* of 31 August 1971 was headed 'Pop festival— should have called police sooner'. This was when 39 Hell's Angels terrorized people at Weeley pop festival in Essex, for love of their motor cycles, as Colchester magistrates were told. Mr Michael What-cott, a motor cyclist, said that the gang had not been violent and had threatened no one. 'Our boys were given permission to take jobs fighting fires. *The only thing we have got in the world are our motor bikes, our fellows and our girls.* We have spent years and every spare penny building up these bikes; they are our one and only life. It hurts to see them damaged, it hurts to see our wounded, but they are used to that and so are our girls.'

This statement indicates that their lives centred upon their motor-bikes and their girls and this is borne out by reading *Buttons: the making of a president* (Mandelkau 1971). The cohesion of the group is their predominant concern and therefore other people are seen as a phenomenon outside the experience of the group, and this leads to frequent confrontations.

One other aspect of the Hell's Angels is their concern for badges and uniforms, which are also seen as extensions of an individual's person. Badges function as a simple method of identifying a group and those outside it, as indeed uniforms do when they assist the definition of social roles. This fixed element helps to define a social position.

Recently a controversy on why children of inner-city areas of the USA show lower academic achievements than the rest of the population has been published in the *Atlantic Monthly* (229, pp. 59–67, 1972). W. Labov described the differences between the standard English of the classroom and the vernacular language used by members of the street culture. His study, like that of Basil Bernstein, is concerned with the relationship of language, culture, and social structure, and shows that the intellectually poor performance of ghetto children is related to the differences of culture and social experience. He produces evidence incidental to his main concern that shows rank-ordered social relations as the dominant structural influence in the form that the culture takes.

> We see many speech events which depend on the competitive exhibition of verbal skills, singing, sounding, toasts, rifting, louding—a whole range of activities in which the individual gains status through his use of language. We see the younger child trying to acquire these skills of older children hanging about the outskirts of the older peer groups and imitating their behaviour.

I am not directly concerned with the nature/nurture controversy, which is wholly wrongly conceived, but evidence is here further provided that peer groups create rank-ordered social relations wherever the culture permits it.

At the beginning we argued that an infrastructure exists, determining the form of social cohesion in sub-human primate groups, and that this takes two forms, the agonistic and the hedonic. This is not the same as saying that in humans non-verbal means of communication take these two forms, but that the cohesive infrastructure, especially the agonistic, is capable of determining not only the form of the social relations but also the consequent dependent forms of communication at any level at which the individual relates himself to others. This has been demonstrated in human society because language has been shown to be subordinate to this infrastructure wherever restricted codes are in operation. We must envisage, therefore, that this same infrastructure will, in the appropriate circumstances, be capable of dominating the form of social relations expressed through convention or the structure of thought. If such occurs, man is not free, but if hedonic-type relations can prevail, the dimensions of man's freedom are as wide as we know they can be.

THE ASSESSMENT OF ATTENTION IN GROUPS

The original concept of attention structure was derived from an attempt to discover the mechanism underlying rank-order in sub-human primates. Having identified the nature of cohesion in the structure of attention, the problem now consists of assessing this element in its own right. One of the first direct pieces of work on this was that of Kummer and Kurt (1963). They showed that, in the hamadryas baboon, the interaction of the members of two adjacent one-male groups (next to each other on a sleeping-niche) was largely confined to other individuals of their own one-male group, despite the fact that many of the individuals were as close to members of the other group as to members of their own. This showed clearly that their attention was primarily confined within the one-male group, and that this also had a large component directed to the male whom they followed as soon as he got up and moved off.

Subsequently, Kummer has carried out a series of observations on pairs of males of different species confined in a large enclosure in their natural habitat. When two members of the same sex of any species are confined in a large enclosure and the distance separating them is assessed at regular intervals, the results can be plotted on a distribution curve with a mean of two to three feet separating the females, indicating that the tendencies attracting the males to each other, and those tending to separate them, are in balance. This is consistent with the way we know their attention is fixed on the dominant individual from whom they keep a mean distance in wild groups. The curves of the distances separating male patas monkeys, on the other hand, are of quite a different kind. Not only is the mean distance about fifteen feet, but the curve of the distribution is not smooth or even normally distributed; it shows a series of peaks separated by troughs which can be duplicated if a second set of observations is made on the same pair or a different pair in the same enclosure. If, however, the objects in the enclosure are distributed in a different way, the peaks and the troughs appear in different parts of the distribution curve, showing that the attention of the male patas monkeys is divided between their companions and prominent objects in the enclosure. When they move closer to their companions they do so only in steps which are determined by the distances separating prominent objects in the enclosure. This is consistent with their known behaviour in the wild, where a patas male, being separated from the females by a long distance, keeps close to prominent objects in the environment where he displays on the approach of predators.

One of the subdivisions of attention structure is therefore between prominent or dominant individuals, on the one hand, and features of the environment on the other. This same division of attention has been demonstrated in Gary Emory's study (Emory 1975) of a gelada baboon colony and a mandrill colony housed in identical enclosures in the San Diego zoo. He used head position and body position in relation to the direction towards or away from the dominant male, and subdivided the segments of a circle into four quadrants, as shown in *Figure 1*.

He showed that the frequency of head and body positions in these segments distinguished the pattern of attention structure in these two species. Members of the gelada colony showed a preponderance of head and body positions in quadrant 1, looking towards the dominant male. In the mandrill colony the members spent a large proportion of their time looking towards the environment and away from the dominant male, with a preponderance of head and body positions in quadrants 2, 3, and 4. Head and body position is therefore one metameter by which attention structure can be studied and within which greater detail and elaboration should be made possible in the future.

Figure 1

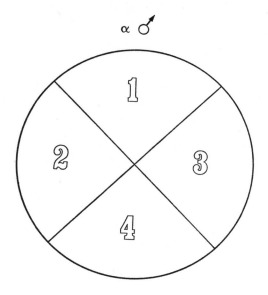

This being so, it could be inferred that head position does give an indication to another observer of whether or not an individual is paying attention to him. An experiment by von Cranach (1971) reports that head position and head movement play a significant part in a human receiver's estimate of whether or not an individual is involved in a 'face reaction', as it is called, with another individual. If this can be further substantiated then head movements constitute a metameter for both man and monkey. In a further study he found that the beginning of verbal communication is usually preceded by an orientation of the body, head, and gaze. So we assume that these behaviours are hierarchically ordered and express the intensity of readiness to communicate. Gaze, head, and body are turned towards the partner, and in inhibited, depressed patients an increase in orientation reaction seems to announce improvement, even before verbal interaction is resumed. Further evidence that a correlation exists between gaze movement and total motoric (that is to say, movements of the rest of the body) is found in the work of R. Jung (1963). He summarizes his findings by saying that, in men, voluntary gaze movements are almost never restricted to the muscles of the eye. Neck and body muscles are involved in auxiliary innervation. Eye movement and nystagmus are but sections of a general orienting movement of the total motoric. Further, since we generally tend to look markedly at a partner (or markedly *not* to look at him), horizontal gaze movements are practically always accompanied by movement of at least the head when interrupting longer periods of looking at the partner or looking away from him.

THE MANIPULATION OF GROUP ATTENTION BY INDIVIDUALS—
ADVERTENCE

Although the concept of attention structure was initially formulated in terms of the preponderance of attention paid by subordinate individuals to those higher in rank, especially to a single dominant individual in an agonistic society, features of the hedonic mode showed quite clearly— from the ways chimpanzees behaved in particular—that each individual to some extent competes for the attention of others in the group by forms of display. A period of mutual display, involving a somewhat competitive form of attention-*seeking* rather than attention-*paying*, was the way that the infrastructure of social cohesion was built up in a chimpanzee society. Studies of our own colony of *Macaca fascicularis* (long-tailed macaques) have shown us quite clearly (Chance and Jones, in prep.) that subordinate individuals solicit the attention of the dominant male, and in doing so enter into a relationship with him which is not previously determined in its character. The act of soliciting therefore appears to be attractive in its own right. From time to time the oldest juvenile male, Arlo, looks in the direction of Percy, the adult male, and, raising an arm above his head, brings it down with a sharp striking movement. He does this at some distance—several feet—from Percy, who often responds with a lunge threat, and may chase him, during which time Arlo retreats. He may repeat the gesture but give a facial expression we call 'mime', which is a signal of appeasement in the circumstances of aroused threat, and hence indicates his willingness, or wish, to approach. This may eventually appease Percy to the extent that he will reciprocate with a 'mime', and then a play session may follow in which Arlo leaps across Percy—just out of his reach—and eventually approaches him, 'miming' vigorously and enters into a wrestling match. They may hug each other ventroventrally. Arlo's success in initiating a play session, however, is not one hundred per cent and he may end up having to escape from Percy's threats. This soliciting of attention is not quite the same thing as calling attention to oneself, as monkeys do when they jolt and shake the branches or jump up and down on a prominent place. This simply brings the attention of everyone on to them and does not involve them with a particular individual. Nevertheless, the relation of Arlo to Percy shows quite clearly that one individual attempts to manipulate the attention of another individual in the group, and that therefore the study of the structure of attention must take this element into account.

Von Cranach (1971) observes that the assumption that 'looking behaviour' functions as a signal can be found in many studies of the subject as an explicit or implicit supposition that determines the design of the investigation and the interpretation of the results. In discussions of this kind, as pointed out earlier, the assumption is that the signal is providing information, whereas in the case of Arlo it appears that the signal is merely attempting to establish a channel of communication. This is the prior underlying condition for establishing states of communication, and yet Argyle (1967) and Kendon (1967) start from the other end when, as Argyle says:

> We may distinguish three ways in which visual orientation functions in interactions. One, to look at another is a social act in itself; two, to meet the gaze of another is a significant event and may often be an important part of the goal sought in interaction, and three, in seeing another, much important information about him may be gathered in addition to his direction of gaze.

Von Cranach (ibid.) places the same emphasis when he says that his own paper will deal with the first two of these statements, which maintain that the gaze of one partner is significant for the behaviour systems of both partners, thus inferring a communicative function. In another place he says, however, that gaze direction and eye movement are used only in combination with other orienting behaviours emphasizing a point we have already seen in the position of the head and body in relation to attention. Analysis of conversation between individuals undoubtedly leads to the view that many of the changes in head position, eye contact, and other components of a non-verbal interchange, are modulating the communication channels of the speaker and listener. So one of the differences between human and non-human communication is that the non-verbal channels in the non-human primate are modulating as well as carrying the message whereas in humans the non-verbal is modulating an interchange or a system of communication that is largely verbal.

The question now arises: what is the real relationship between these two systems? This was the object of a collaborative study between Hilary Callan, an anthropologist, myself, and Tom Pitcairn, a biologist (Chance, Callan, and Pitcairn 1973). In order to do this we realized we would have to go back a stage in order to avoid being clamped into the framework of thought that governs existing procedures for the study of this relationship. We soon realized that we had to study the microstructure, which we defined, for the time being, very loosely, as the total of all that is patterned and systematic about social interaction in the immediate and quasi-immediate presence of others. At the start we took 'presence' to mean straightforwardly 'physical location', and by 'small-scale social activity' we meant social behaviour involving small numbers of individuals rather than broad categories or groups of our society. Nevertheless, we were dealing with the study of human groups, so we allowed ourselves to be guided by Goffman (1969), who defined the study of social microstructures as the bringing into one framework of concepts and findings derived from three areas of inquiry: the individual personality, social interaction, and society. By attempting to assess, or be aware of, the influence of society, we refer to everything a socially functioning human being does in face-to-face interactions which is done partly with reference to a submerged series of ought-statements derived from culture, learning, the person's role, the nature of the situation, and so forth. In this study we are concerned with two of the three components indicated by Goffman, namely social interaction and society.

We chose to study the inmates of a psychiatric hospital for two reasons: first, the rules of the institution were relatively simple and easy to identify, and, second, in another study (Grant 1969), mental patients had exhibited patterns of social behaviour strikingly similar to those of sub-human primates. Our observations were made on a schizophrenic group and a psychotherapeutic group. Before we began, we found it necessary to distinguish a response in the psychological sense as an elicited piece of behaviour and also in the social sense as an acknowledgement. These were, so to say, at two levels, and yet closely interrelated in their action. The fact that an acknowledgement was possible in a group, or need not be accorded by the group, made it clear that at least part of the behaviour of the patients was an attempt to become part of a group by being accepted into it. In fact, social cohesion had to become social acceptance in a human group in order for the group to operate.

An acknowledgement means acceptance by the group. In attentional terms this means manipulating the attention of the rest of the group in such a way that the individual becomes recognized and takes his place as a participant in the ongoing activity of the group, whatever that is. We have seen from Kendon (1967) and Argyle (1967) that non-verbal components are used in these circumstances to modulate the conduct of the conversation. What became evident from our study of the psychotherapeutic group was that many individuals had not yet reached this position of acceptance and were constantly striving, apparently in a partially familiar situation, to gain acceptance. *Advertence* was the name we gave to the process by which they advertised their presence and their willingness to become part of the social activity. (Activity is here defined as the purpose of the group as well as the actual ongoing interaction between members.) In our observations of the psychotherapeutic group the aim was to identify the level of behavioural and social integration at which advertence operated. It eventually became clear that there were in fact four rough types of self-presentation, which we termed *alert*, *away*, *closed*, and *huddled*. To various degrees, and in this order, these forms of advertence constituted more and more exaggerated forms of self-presentation, representing, curiously enough, the greater degree of withdrawal from participation in group activities as the patients moved away from *alert* through *closed* to *huddled*. In fact, it was only those patients showing *alert* and/or *away* from time to time who participated in what was then the activity of the group, namely, discussing the problems presented by an individual who was at that moment the centre of interest and concern. Hence participation in this sphere of social activity indicated an ability to handle information that was not directly concerned with the individual's own problems. Patients who had failed to present themselves and their own problems in a sufficiently acceptable way to the group were driven into, or acquiesced in, the role played by those in *closed* and *huddled*, two rather more fixed categories of advertence. This general pattern is repeated in the forms of advertence identified in the schizophrenic group, with different elements. Here, the location of patients at marked distances from one another, the patrolling of these intervals of physical space, and the use of certain idiosyncratic displays—usually involving the whole body— which were specific to the individuals concerned, were prominent advertive techniques. In a less pronounced way, so was the use of physical objects and features of the environment such as tables, chairs, newspapers, and cigarettes. Finally, institutional factors such as hospital routine were combined with the latter to achieve, avoid, and modify social contact.

In this group there were usually two or three individuals occupied in conversation and continuing activities of one kind or another, usually revolving round the routine of the hospital, such as serving tea and meals and clearing up after other activities. They kept up a continuous interest focused on the nature of the whole proceedings. Away at the other end were individuals who were spatially isolated and who had to adopt various exaggerated forms of display in order to obtain attention for a brief interval during which contact was made and then rapidly lost with other members of the group. (This reminds one of the exaggerated displays of chimpanzees, indulged in when two groups meet after having been separated from each other for some time.) Mutual display and mutual recognition involve gross movements of the whole body. What appears to be characteristic of advertence is that it is used as a means of establishing group acceptance of an individual by manipulating the attention of the group. Hence the initiation of a new group so that the individual can play a socially accepted role, or be personally accepted into the activities of a group when these develop, are two forms of such intervention. Advertence, therefore, is a form of interaction of group members involving the reaction to group relations, and is typical of at least semi-unfamiliar conditions.

Another term is needed for the part played by non-verbal elements in the communication system of groups where coherence has been established at the level of operational activities. In such groups, the speaker, for example, announces by eye flash that he is going to start a new point, or emphasizes one in the same way. The audience, by paying attention and being in *alert*, recognizes his attempts and urges on his participation. When individuals are seen to be 'sleeping' in the audience, it is another way of saying that quite a few of them have gone into *huddle*, and that the speaker's activities no longer matter to them or can be taken for granted by the group acting as a whole. In groups where technical discussions are under way there is a matrix of common understanding of a reality principle which allows an individual to come in with his contribution at the point where the technical requirements involve the presentation of the knowledge which he has in his specialized role. If he participates in another form it may be an attempt to direct the activity of the group into different channels, but this can only be done with a certain amount of advertised opposition; that is to say, withdrawing the attention of the group as a whole from the current form of ongoing activity in an attempt first to transfer the attention of the group to himself, but ultimately to get a realignment of the group's discussion.

One can now see that there are various degrees of coherence possible in types of discussion dependent partly on the compatibility of the information flow from different individuals towards the common pool. These may be graded from negotiation, through forms of bargaining, to various types of international confrontation. It is proposed that the term *annunciation*, or the annunciative role, should be used for the way personal advertising is used or attention manipulated in a group with fully operational designation. 'Advertence' could then be reserved for defining the manipulation of group attention where it is used mainly for the acceptance of an individual within an existing group. Charisma then becomes a special form of advertence used by individuals to form a new group, whether, as Max Weber (1968) contends, the group is formed to promote a new philosophical, religious, or ethical point of view, or does not go beyond placing the individual at the focus of a group's attention.

The essence of understanding here lies in being able to detect whether the advertive activity of an individual is directed towards gaining for the group an acceptance of a principle or mental concept of its social role, rather than merely establishing a new relationship of the members of the group to the advertiser. Advertence may place the advertiser at the focus of attention of the members of the group or merely establish the individual among the accepted communicants.

The distrust of charismatic activity by bureaucrats is derived as much from their resistance to the simple realignment of the group activities as from the dangers that arise in an individual achieving a focal position in the attention of the group merely to bring about the disruption of its activities without a constructive alternative. This latter is a real possibility, for it is a phylogenetically old development of such a state of affairs that such an individual can fix his focal position by the use of threat, whether consciously or because of phyletic compulsions. For when this happens, revolt is the ultimate phyletic sanction, providing the only way of opening up the group to future change. As yet, the conscious formulation of a counter to agonistic charisma, let alone tyranny, has not yet been found and is urgently required.

111

M. R. A. Chance

In agonistic forms of social cohesion the whole group is involved in the common activity of paying attention to a focus to avoid being aggressed against, but attention in the hedonic mode does not *ipso facto* provide social cohesion for the group as a whole. When individuals attempt to gain entrance into a group, or when an individual creates a group by drawing the attention of a number of individuals to himself, advertence is the method of achieving the attachment of the individual to the rest of the group.

Bibliography

ANGST, W. 1974. *Das Ausdrucksverhalten des Gavaneraffen* Macaca Fascicularis, *Raffles 1821.* Berlin and Hamburg: Paul Parey.

ARGYLE, M. 1967. *Psychology of Interpersonal Behaviour.* Harmondsworth: Penguin Books.

—— 1972. Non-verbal communication in human social interaction. In: R. A. Hinde (ed.), *Non-verbal Communication.* Oxford: Oxford University Press.

BERNSTEIN, B. 1965. A socio-linguistic approach to social learning. In: J. Gould (ed.), *Penguin Survey of the Social Sciences.* Harmondsworth: Penguin Books.

BOLWIG, N. 1963. Observations on the mental and manipulative abilities of a captive baboon (*Papio droguera*). *Behaviour* 22 (1): 24–40.

CHANCE, M. R. A. 1960. Köhler's chimpanzees—how did they perform? *Man* o. s. 60: 130–135.

—— 1967. Attention structure as the basis of primate rank orders. *Man* n. s. 2 (4): 503–518.

CHANCE, M. R. A., CALLAN, H. M. W., & PITCAIRN, T. K. 1973. Attention and advertence in human groups. *Social Science Information* 12 (2): 27–41.

CHANCE, M. R. A., & JOLLY, C. J. 1970. *Social Groups of Monkeys, Apes and Men.* London: Cape; New York: Dutton, 1971.

CHANCE, M. R. A., & JONES, E. In preparation. Paternalistic behaviour in the long-tailed macaque. MS.

CRANACH, M. VON 1971. Orienting behaviour in human interaction. In: A. Esser (ed.), *Behaviour and Environment: The Use of Space in Animals and Man.* New York and London: Plenum Press.

EKMAN, P. 1972. Universal and cultural differences in facial expressions of emotion. In: J. Cole (ed.), *Nebraska Symposium on Motivation.* Nebraska: University of Nebraska Press.

EMORY, G. R. 1975. Attention structures as a determinant of social organisation in the mandril (*Mandrillus sphinx*) and the gelada baboon (*Theropithecus gelada*). In: M. R. A. Chance & R. R. Larsen (eds.), *The Structure of Social Attention.* London and New York: Wiley.

FOX, R. 1967. In the beginning: aspects of hominid behavioural evolution. *Man* n.s. 2 (3): 415–433.

—— 1972. Alliance and constraint: sexual selection in the evolution of human kinship systems. In: B. Campbell (ed.), *Sexual Selection and the Descent of Man, 1871–1971.* Chicago: Aldine.

GOFFMAN, E. 1969. *The Presentation of Self in Everyday Life.* London: Allen Lane.

GRANT, E. C. 1969. Human facial expression. *Man* n. s. 4 (4): 525–536.

JENSEN, A. R. 1968. Patterns of mental ability and socio-economic status. *Proc. Nat. Acad. Sci.* 60: 1330–1337.

JUNG, R. 1963. Neurophysiologische Untersuchung. Methoden: Nystamographie. In: von Bergman *et al.* (eds.), *Handbuch der Inneren Medizin E-D.* Vol. 1. Berlin: Springer.

KENDON, A. 1967. Some functions of gaze direction in human social interaction. *Acta Psychologica* **26**: 22–63.

KUMMER, H. 1968. Social organization of hamadryas baboons—a field study. *Bibliotheca Primatalogica* No. 6. Basel and New York: S. Karger.

KUMMER, H., & KURT, F. 1963. Social units of a free-living population of hamadryas baboons. *Folia Primat.* **1**: 14–19.

KÖHLER, W. 1927. *The Mentality of Apes.* London: Methuen.

MANDELKAU, J. 1971. *Buttons: the Making of a President.* London: Sphere Books.

MASON, W. A. 1965. Determinants of social behaviour in young chimpanzees. In: A. M. Schrier, H. F. Harlow, & F. Stollnitz (eds.), *Behaviour of Non-human Primates.* Vol. 2. New York and London: Academic Press.

MENZEL, E. W. 1972. Spontaneous invention of ladders in a group of young chimpanzees. *Folia Primat.* **17**: 87–106.

REYNOLDS, V. 1973. Man also behaves. In: J. Benthall (ed.), *The Limits of Human Nature.* London: Longmans.

REYNOLDS, V., & LUSCOMBE, G. 1969. Chimpanzee rank order and the function of display. In: C. R. Carpenter (ed.), *Second Conference of the International Primatological Society: Behaviour.* Vol. 1. Basel and New York: S. Karger.

REYNOLDS, V., & REYNOLDS, F. 1965. Chimpanzees of the Budongo Forest. In: I. DeVore (ed.), *Primate Behavior.* New York: Holt, Rinehart and Winston.

SADE, D. S. 1965. Some aspects of parent-offspring and sibling relations in a group of rhesus monkeys. *Amer. J. Phys. Anthrop.* **23**: 1–17.

SCHALLER, G. B. 1963. *The Mountain Gorilla: Ecology and Behavior.* Chicago: University of Chicago Press.

SCHILLER, P. H. 1957. Innate motor action as a basis of learning. In: C. H. Schiller (ed.), *Instinctive Behavior.* New York: International Universities Press.

TIGER, L., & FOX, R. 1971. *The Imperial Animal.* New York: Holt, Rinehart and Winston; London: Secker & Warburg.

VIRGO, H. B., & WATERHOUSE, M. J. 1969. The emergence of attention structure among rhesus macaques. *Man* n. s. **4** (1): 85–93.

WEBER, M. 1968. *Max Weber on Charisma and Institution Building.* S. N. Eisenstadt (ed.). Chicago: University of Chicago Press.

Lionel Tiger

Somatic Factors and Social Behaviour

Any effort to demystify the causes and nature of human behaviour logically must scrutinize what human organisms bring to social behaviour and how their internal secretions and states are affected by it. Unless it is assumed that the body is a neutral unit in behavioural transactions—which *in extremis* an understanding of neurological and other organic lesions renders an untenable working principle—then it becomes useful for students of explicitly social behaviour to attend to recent work in the sciences that seeks to gather information about the human body and its functioning. Such a review becomes essential when, in an interesting turnabout of the nature–nurture controversy, claims are made about the effect of social process on organic conditions; for example, the occurrence of the 'placebo effect', and an evident synchrony of menstrual cycles among human females as a function of small-group behaviour and friendship patterns. The purpose of this contribution is to propose that an effective functional unity is demonstrated, or at least strongly suggested, by data which purport to relate somatic states to social states, and that this compels insertion or consideration of somatic processes and adjunctive ones in investigating even macro-social behaviour. Doing this raises numerous vexatious problems the solution of which is not necessarily implicit in otherwise successful anthropological and sociological techniques. But this is reason enough to hasten rather than delay the effort against some cumulatively epiphanous event which will of itself resolve the technical and theoretical problems of integrating natural and social science. In so far as natural scientists continue to produce considerable quantities of data and theories which at least on the face of them may bear relationship to the work of social scientists, then an assimilative rather than a restrictive attitude becomes essential. Otherwise social scientists will be made scientifically obsolete by their commitment to work with an unnecessary 'black box' assumption about human biology and 'human nature'.

In this paper I wish to review some relatively straightforward findings from the natural and physiological sciences that offer evidence about the porousness of the boundary between the domains they customarily investigate and those of the social sciences. My intention is not primarily to advance substantive comments about social reality or nature, nor to examine extant propositions about the causation and process of behaviour. Rather, it is to discuss the process and assumptions that pro-duce such propositions and how they may be informed by the work of the biological sciences. On this formal note, it may be appropriate to anticipate the 'Durkheimian' objection to such an essay (Gluckman 1972) that it is reductionist—the word being used in its pejorative sense of oversimplifying, trivializing, 'turning the sociologicial into the bio-logical' and the multifactorial into the monocausal. This is an undeni-able possibility which I intend to avoid, if only because of a methodo-logical assumption that any physiological factor in social behaviour can be only that—one factor among a number of others. In fact, rather than such an approach stimulating an oversimplification, the very oppo-site is decidedly more likely: that trying to understand the biology as well as the sociology of behaviour will become overcomplicated and that attention to the fine-grained responses will preclude or cloud a grasp of overall social structures. However, both these dangers may be miti-gated by adopting the working principle of modern behavioural biology: that any action a complex, gregarious animal undertakes must involve a range of levels of integration, kinds of process, and intensities and duration of response (Tinbergen 1963; 1972). Furthermore, the demon-strably sturdy procedures of synthetic and comparative biology rest on an appreciation of both endogenous and exogenous factors in the appearance of behaviour (Ghiselin 1972; see also Ghiselin 1969) and, more particularly on what profiles of learning/predisposition charac-terize individual species (Seligman and Hager 1972). As the psychiatrist Hamburg has reformulated the nature–nurture problem, what must be determined is not simply what is learned and what is automatic, but what animals find easy to learn when it is taught them (e.g. speech among humans) and what is difficult (e.g. speech among other primates, and sphincter control in all, including humans). There are obviously certain limits to this; for example, Eibl-Eibesfeldt has shown that con-genitally blind children display species-specific facial expressions even though they have never seen them as stimuli to model (Eibl-Eibesfeldt 1970). Also, it has been demonstrated recently that newborn monkeys whose nervous systems have been deafferentated and whose sight has been withheld nonetheless develop most types of movements of the forelimbs, and with only one or two weeks' delay in comparison with normal monkey infants (Taub, Perrella, and Barro 1973). Even in the Soviet Union (where behavioural science has been substantially circum-scribed by systems of explanation based on Pavlovian psychology of the stimulus–response genre), there are recent indications of an enhanced concern with ethological forms of explanation. Though for institutional or tactical reasons these are described as ecological rather than etho-logical, nonetheless, according to Kovach, the concerns and findings of the papers which are increasingly published are essentially involved with the connection between the evolution of an animal, its genetic endowment, and its contemporary behaviour under a variety of circumstances (Kovach 1973). Perhaps stimulated by that country's assertive success in microbiological science, even France's primarily structuralist–intellectual tradition of sociological explanation has been recently augmented by a strong statement of the interpenetration of even cognitive with genetic systems, to say nothing of other more obvious ones such as food-gathering and sexual systems (Morin 1973). At the straightforward physical level, White and Parsons have affirmed that the linguistic and other socio-cultural differences commonly described between Aborigines of western and eastern Arnhem Land are also accompanied by genetically-based dermatological differences, thus attesting to these groups' distinctiveness as breeding groups as well as social ones (White and Parsons 1973). In sum, the implication here is not only that the nature–culture boundary is porous—by now a trivial comment—but that what happens along it, and on one side with reference to the other, are amenable to investigation and disconfirmable analysis.

The body is one unit among others in the system of behaviour. It is both an envelope in which certain responses are contained (fear, stress, rage, etc.) and also an energizing source of a variety of kinds of stimuli with social consequences—sounds, visual gestures, smells, rates of movement, etc; aspects of the system involving the body are common in the study of both humans and the other animals (Argyle 1972). Perhaps the most elaborate review of the general area of investigation along these lines from the anthropological perspective is Chapple's (1970). He focuses his survey of biological factors in behaviour on a variety of specialized subjects which combine to form a reasonably comprehensive overview of significant areas of research, e.g. primate studies; ethology; behaviour genetics; circadian rhythms; territoriality; the relationship of physiological process to emotional experience; the means and ends of measuring social interactions; the comparative psychology of patterns of learning and forgetting; the physiology of the senses and the psychological consequences of their uses; and the biology of language. In each of these foci of work, Chapple outlines the findings which in his opinion must have effects on the perception and analysis of human social behaviour. For example, he cites Helbrugge's findings about circadian rhythms which suggest, at least among under-15-year-olds, a physiological tendency towards a siesta between 1 p.m. and 3 p.m. Further: 'Helbrugge's finding that ability to concentrate in eleven-year-old children is remarkably reduced before 8 a.m., between 2 p.m. and 4 p.m. and after 8 p.m. has interesting educational portents.' A comprehensive review of materials up to 1970 pertaining to the circadian rhythms Chapple discusses is Luce's *Biological Rhythms in Psychiatry and Medicine* (1970). Of particular interest to students of simple societies is Luce's indication that technologically complex communities inadequately attend to such rhythms. A useful source of information about human biology could well reside precisely in the percep-

117

tions of time and bodily functions represented by the practices of members of simpler societies: an ethnobiology, in fact. The value of gathering information about the patterns of internal cyclic conditions such as digestion, temperature, and metabolism is indicated quite clearly by such work; and the classical ethnographic ideal of carrying on a study over a year's cycle finds its justification not only in the obvious variety of behaviours observable over this period, but also in terms of the less-known and more subtle alterations by which the human animal is marked over time, and which have only very recently become clearer as phenomena to scientists. For example, while it is a reiteration of the ethnographically obvious, Ismail *et al.* (1972) note that human sexual activity clusters in the night-time, but that it is also negatively correlated with levels of plasma cortisol. Gordon and Bernstein (1973) have indicated that among the other primates there is in rhesus monkeys a clear annual cycle of sexual behaviour and suggest that control of this cycle is through seasonal variables acting upon the females only to alter their endocrine status; even without direct physical contact, females can communicate their endocrine status to males. The importance, for even general theories of social structure, of noting circadian rhythms is underlined by Vessey's (1973) recent report of his night-time observations of free-ranging rhesus monkeys. Using an image-intensifier for observations, this pioneering study has revealed *inter alia* that 33 per cent of sleeping clusters were mother–infant pairs; that 63 per cent of sleeping clusters were made up of maternally-related monkeys (this is of quite crucial importance to Fox's argument in this volume); that 17 per cent of sleeping clusters were solitaires; that 21 per cent of sleeping clusters contained at least one non-related member; and that clusters with non-related members were more common during the breeding season—a form of gregarious permissiveness presumably under endocrinological control, following Gordon and Bernstein. As for daytime grouping, McGuire *et al.* have indicated the importance of understanding the daily time-dependent round of movement of primate troops in order to become aware of the extent to which their intergroup behaviour is a function of such movement, and of how elaborately and with what structure the animals govern their daily activities in the context of a large-scale gestalt of troop movements (M. T. McGuire 1974). Bernstein (1972) has also described the rhythmic pattern of monkey social behaviour as a function of temporal climatic changes.

While there has been a fairly lively ongoing interest in the relationship of childbirth and season (e.g. Black 1973; Kanebar and Mukerjee 1972), there has been somewhat less commitment of interest to pursuing possible connections between climatic/geographical factors and ritual and other systematic behaviours. This may well be informed helpfully by the burgeoning body of biological information. Again, what is implicit here is the addition of another factor to the research mixture, not the replacement of already familiar ones. For example, one recent study that constructively employs physiological as well as traditional anthropological methods of gathering data is that by Foulks of arctic hysterias among North Alaskan Eskimos (1972). What Foulks (an M.D., incidentally) has shown is that during certain times of the northern arctic year, a physiological susceptibility to hysteria exists (with real concomitants in neural transmission systems), that it can be readily triggered psychogenically by the routine physiology of a sudden bout of anxiety, and that traditional Eskimo culture exerts, through its work values and expectations, an unduly large amount of 'stressor'-type pressure on its individual carriers. In effect, Foulks has sought to dovetail a demonstrated hyperirritability of key central nervous system structures and operations with a cultural framework that he concludes is hypersuggestive of stimuli to extreme anxiety. In a similiar kind of analysis Bolton (1973) has argued that the very high rate of violence among the Quolla is supported by conditions of general hypoglycaemia as well as by the obvious sociogenic factors open to inspection.

SEX ROLES AND INTERNAL SECRETIONS

Two major systems promote and integrate bodily functions in animals —the nervous and the endocrine. Traditionally, these have been investigated rather independently of each other. However, there is now an accumulating file of data and their synthesis which indicate the extent of the interdependence of these systems. In this context I want to reflect some recent work that focuses on one of the clearest cases of biological infrastructuring of bodily function—sex differences.

During the past four or five years a paradoxical research situation has emerged which has affected the intensity and equability of scientific discussion. Because of ardent and incisive challenges to existing attitudes to females and to their social conditions in a variety of communities, a large literature is forming which coalesces around the hypothesis that the most effective cause of behavioural differences between human males and females resides in the procedures of socialization and in different or frankly invidious life-chances which are available to females (e.g. Holter 1970; Figes 1970; Epstein 1970; Millett 1970). The merits of the political, economic, and ethical aspects of this argument are plain. Equally apparent is the value of the feminist critique of social scientists when it focuses on the heretofore relatively scant attention to sex roles and on the tendency for the category 'man' or 'person' to reflect implicitly or patently male-oriented preoccupations. (See my own comments on this, with particular reference to political scientists and sociologists, Tiger 1969; see also Dearden, 1973.) As Corinne Hutt has introduced the problem in her review of the literature in this field, *Males and Females* (1972a):

In recent years there has been a resurgence of interest in the topic

119

of sex differences. . . . The recent interest . . . is no doubt due to the efforts of certain movements directed at social change. An unfortunate consequence of this . . . seems to be the tendency to refer to such differences, only to dismiss them as cultural excrescences or to demonstrate the creation of male and female stereotypes by society. Having met, repeatedly, with rejections of any suggestion, however covert, that some human sex differences may be of biological origin, it seemed important to set the 'sexual differentiation' story in perspective. . . . In much developmental and social psychological writing too little cognizance is taken of the structure and function of the brain, much less of the constraints set by the nature of its organization. I make no apology, therefore, for stating the case for the biological bases of psychological sex differences.

Hutt's and other similiarly concerned analyses are chiefly attentive to psychological and physiological behaviour and rather less to macrostructural activity (e.g. Bardwick 1971; Garai and Scheinfeld 1968; Gray 1972. Van den Berghe 1973 is an exception). Furthermore, it is clear that with respect to a great variety of factors—size, speed, strength, spatial ability, verbal skill, etc.—male and female curves of response overlap more or less considerably and the similarities may be in a statistical sense more marked than the differences. (For example, see Wilmore 1973.) However, precisely because of this amount of overlap it becomes important to assess and understand the differences if only to seek accurate predictions about options and constraints in maintaining societies or changing them. In addition, attention to sex differences among students of the social behaviour of other animals, the primates in particular, has been immensely important in stimulating theories about the relationship between social behaviour, genetic status and change, selective patterns, and functional and dysfunctional qualities of particular socio-economic patterns. (For a recent review see Campbell, ed., 1972; also Crook 1973.) While the human case is more complex in virtually every way, Occam's razor lurks near the corpus of any theory that systematically and even assertively excludes such information as is available about mammals and the primates, and the infrastructural somatic factors humans share with other members of these orders, as well as those we do not share.

A paper by Washburn and McCown (1973) underlines the hazards of an attention to sex differences that does not begin with a concern for function as well as form. As part of a review of the implication of physical sex differences in the primates, they begin not with the conventional assumption that these differences must be expressed in male-centred terms, as in, for example, 'the Gibbon female is 90 per cent of the male in size, weight, etc.'. Rather, they begin with the assumption that the basic primate anatomy is female and that, apart from the specific reproductive organs, what is added onto the males is the apparatus for aggressive behaviour. The female establishes the basic size of the species, its adaptation to diet, etc., while the male adds

> the anatomy of bluff and fighting (collectively, aggression). Contrary to our cultural biases, very little anatomy is directly concerned with sex, and a great deal with the ability to fight . . . [e.g. the] . . . gorilla with minimum genitalia and enormous differences in size, muscles, teeth, etc. This view of the male–female differences allows one to see behavioural–structural patterns so that the relations of play, hormones, muscles, teeth, callosities, hair, etc., *all* become understandable (Washburn, pers. comm. 1973).

This technique of analysis shows that, while human sex differences may be in general less exaggerated than those of, say, baboons or chimpanzees, with respect to the apparatus of aggression human males have evolved roughly twice as much equipment as females. This specific differentiation of 200 per cent, then, does not repeat the general value of sex differences which is closer to 85 per cent. Thus focusing on females rather than males in these anatomical ways offers the possibility of an improved evaluation of the roles of the sex differences in behaviour which have yielded demonstrable species-specific consequences for humans as well as for other primates. This in turn facilitates an assessment of the biological and physiological infrastructure of sociosexual roles which all humans share by virtue of their membership of our species.

In effect, to try to assess the causation and function of sex differences without also understanding the phylogenetic and genetic aspects of these differences is to rely with undue confidence on the evidence of the contemporary without adequate historical and prehistorical comprehension. Needless to say, an enthusiasm for evolutionary analysis need not necessarily lead to scientific positions which are stable and unassailable; as Washburn and Ciochon (1973) have noted, 'The history of theories of human evolution is filled with examples of competent scientists, with access to the same facts, drawing different conclusions, and then becoming involved in acrimonious debate often lasting for years'. None the less, at least such debate as emerges will be informed by the best available information about real processes even if its interpretation produces differences of view.

THE ENDOCRINES

An intriguing point of contact between work on the nervous and endocrine systems has been in neurophysiology, which has yielded some understanding of the role of the sex hormones in brain formation and theory: 'The single fact, *that some part of the brain is characteristically different in males and females*, is one of the most significant findings in neuroendocrinology' (Hutt 1972b). As Levine (1966) and Money and

Ehrhardt (1972), among others, have shown, the effect of sex hormones *in utero* is crucial for the establishment of either male or female neurophysiological orientations. Furthermore, as a result of pharmacological errors of dosage and medication, pregnant females given unusually large amounts of sex hormones have borne boys and girls decisively and permanently affected by the drug-stimulated orientation. While these effects may be interpreted, struggled against, distorted, or significantly overcome, Money and Ehrhardt's assertion remains: that the predispositions that intrude into a social situation because of the actors' sexual natures are real phenomena which must be seen as consequential; a synthetic approach is essential.

> On the theory of psychosexual differentiation, it is now outmoded to juxtapose nature versus nurture, the genetic versus the environmental, the innate versus the acquired, the biological versus the psychological, or the instinctive versus the learned. Modern genetic theory avoids these antiquated dichotomies, and postulates a genetic norm of reaction which for its proper expression requires phyletically prescribed environmental boundaries. If these boundaries are either too constricted, or too diffuse, then the environment is lethal, and the genetic code cannot express itself, for the cells carrying it are nonviable.
>
> The basic proposition should not be a dichotomization of genetics and environment, but their interaction. Interactionism as applied to the differentiation of gender identity can be best expressed by using the concept of a program. There are phyletically written parts of the program. They exert their determining influence particularly before birth and leave a permanent imprimatur. . . . Even at that early time . . . the phyletic program may be altered by idiosyncracies of personal history such as the loss or gain of a chromosome during cell division, a deficiency or excess of maternal hormone, viral invasion, intrauterine trauma, nutritional deficiency or toxicity, and so forth (Money and Ehrhardt, 1972).

Of particular theoretical interest to social anthropologists must be the contents of the phyletically written 'program' and what are the 'phyletically prescribed environmental boundaries'. This is in a real sense another version of the traditional quest for 'universals' in human societies, or functional prerequisites. However, to the extent the enterprise can depend upon verifiable and cross-culturally applicable statements about human propensities, an augmented precision becomes possible that is unavailable to those focusing solely on sociogenic processes.

One example of such augmented precision emerges from the considerable amount of information newly available about the effects of hormone secretions and hormone levels on social behaviour, and vice versa. Among some mammals, such as rats, rabbits, and the other primates, a relatively clear picture emerges of the relationship between the largely male sex hormone testosterone and what is defined as aggressive behaviour (Rothballer 1967; Davis 1970; for a more general discussion see Gottier 1972; Gray 1971). Injection of testosterone into females stimulates them to behaviour that more correctly characterizes undrugged males than undrugged females of the species in question, e.g. rats, rhesus monkeys, rabbits. While the situation is less clear among humans, the extreme case of the castrati suggests the social and behavioural importance of withdrawal of male hormones, particularly when castration occurs prepubertally; comparable effects of ovari-ectomy are noted among premenopausal females (Money and Ehrhardt 1972). For the non-human primates, there are some experimental results displaying a connection between testosterone and aggressive behaviour in male rhesus monkeys (Rose and Holaday 1971). This finding has been replicated by Kling of Rutgers Medical School (person. comm. 1972) for both male rhesus monkeys and stumptail macaques.

It was an important finding of students of psychosomatic medicine that individuals' personal circumstances could affect their physical health. Among humans there are indications that particular forms of social organization may have measurable effects on internal secretions. In a group of male officer candidates in the U.S. Navy, Kreuz *et al.* (1972) discovered a reduction in testosterone levels during the initial phases of cadets' training when stress was high and formal harassment and structured humiliation were extremely intense (Kreuz, Rose and Jennings 1972). These levels increased towards graduation when the pressures upon cadets were mitigated by the educational institution and when they were finally converted into supportive and enhancing forms of treatment. A comparable finding of decreased plasma testosterone was made by Rose *et al.* (1972), who introduced formerly dominant (high testosterone level) rhesus monkey males into an established group of rhesus males, who rejected the newcomers' claims to dominance. Among humans again, Kedenburg *et al.* (1974) found a positive correla-tion between recovery rates among a group of psychiatric patients and increasing rates of testosterone secretion; mental health was not solely a factor of cognitive and social circumstances but was also endocrinal. Among rhesus monkeys, the sexually dimorphic character of social activity becomes clear early on in the first year of life. Males are phy-sically tougher than females and reach higher positions in dominance orders. Joslyn (1973) has been able to reverse this experimentally. In a group of three male and three female rhesus newborns all animals were put together, untreated by hormones, for an eight-week period during which time the males displayed the predictable male behaviour and the females the corresponding set for females. However, when females were injected with testosterone propionate they proceeded to act in a more aggressive manner and after several months occupied two of the three predominant positions in the groups. During the androgenized phase, females gained unusual weight and unusual muscle mass. However 'it is unclear whether the increased social dominance in the infant female monkeys was caused by the effects of androgen on the central nervous system, peripheral structures, or both' (ibid.). The male-like behaviour persisted for a year after the last hormone injection. 'The male hormone may have directly induced a permanent change in the nervous system; a more plausible explanation . . . is that social dominant behaviour was so well learned during hormone treatment that it became independent of hormonal support.' Despite the increased aggressiveness of the females, their playful and sexual behaviour has remained feminine (ibid.). What is interesting here is how the physiological state of the animals permitted them to learn appropriate and defensible social behaviours.

In a study of humans which describes a similar pattern Kreuz and Rose (1972) have reported that, among a population of prison inmates, those with histories of more violent and aggressive crimes in adolescence had significantly higher levels of testosterone than prisoners lacking such a history; their hypothesis is 'that within a population that is predisposed by virtue of social factors to develop anti-social behaviours, levels of testosterone may be an important additional factor in placing individuals at risk to commit more aggressive crimes in adolescence'. Their stress on the social context is central here. Their findings suggest that it is not necessarily the case that high levels of testosterone coerce unremitting violent performance—the subjects of this experiment were not themselves violent in prison—but that an internal secretion may be a significant predisposing or catalytic factor under appropriate social circumstances for the display of violent behaviour. In so far as individual testosterone levels appear to be relatively stable over six months at least, then it is possible that particular individuals are more prone than others to commit violent acts, other things being equal: 'It is our suggestion that the testosterone acts in a permissive fashion for the appearance of forms of aggression such as anti-social behaviour. More specifically, testosterone may serve to stimulate increased activity, drive or assertiveness, and in certain individuals this may be utilized in anti-social aggressive acts' (Kreuz and Rose, 1972).

This phenomenon is particularly marked among adolescent males. Understanding its dimension may help to explain the disproportionately bellicose, criminal, and self-destructive behaviour of young human males as compared with females. This may also offer a perspective within which some traditional explanations of adolescence as a sociogenic phenomenon may be seen as unnecessarily limited; for example, the conventional Meadian notion of adolescence as a situation of role change rather than a severely disruptive biologically programmed aspect of the life-cycle.

Mead has recently commented that her emphasis in the nineteen-twenties and thirties was in part an undue reaction against the relatively simple-minded instinctivist theories of the early part of the century. That adolescent-like patterns appear to mark other primate males as well as human, and that both humans and other primates share the rapid increase in testosterone level, lends some strength to an inclusion on methodological and logical grounds of data that do not normally emerge from the social sciences as classically conceived.

For example, David Hamburg has reported that adolescent human females roughly double their testosterone secretions, from a lower base than the males', while the males increase plasma testosterone levels by at least twenty to thirty times, and possibly as much as one hundred times (Hamburg 1972). An earlier estimate which employed less sensitive means of assay estimated the magnitude of increase at ten (Hamburg 1971). The relationship between aggression and young males has also been reported on by Persky *et al.* (1971). Of course, the existence of sex differences may not depend solely or possibly even chiefly on the endocrine systems. For example, Kimura has described sex differences in brain hemispheric dominance that may underlie the frequent finding that males are on balance superior to females in perceptual skills and that females are on balance superior to males in speech skills (Kimura 1973). This results from the fact that the left hemisphere —a female advantage—broadly supports speech while the right—a male advantage—broadly underlies perception. While these are, again, matters of overlapping curves of responses, nonetheless it becomes useful to know what may be the infrasocial components of perceived differences between male and female groups in the population; for example as described in schoolchildren by Lester, Dudek, and Muir (1972), and in general by Buffery and Gray (1972); Dawson (1972); and Fairweather and S. J. Hutt (1972); see also Corinne Hutt (1972b). That endocrine patterns may also affect perceptual skills is suggested by Diamond, Diamond, and Mast (1972), who report a strong relationship between visual acuity and phases of the menstrual cycle, and by Mykytowycz (1972), who describes a relationship between internal reproductive states and the ability to perceive odours.

SUICIDE AND OTHER DISRUPTIONS

Somatic alterations may underlie the aggressive and self-assertive and/ or deviant actions that have been noted above. While a complete relationship between automobile collisions and aggressive activity is of course impossible to establish, accident rates among new drivers who are also post-adolescent are generally high; in a recent report, Richard Zylman (1972) has indicated that, 'Age is more important than alcohol in the collision involvement of young and old drivers.' In effect, the ingestion of a disruptive substance, alcohol, is less likely to stimulate a collision than the youthful age of a male driver; a few beers on top of this evident tendency are disproportionately successful in creating collisions. In this respect, 'drivers 75 and over . . . are very similar to teenagers'.

Some commentators have identified dangerous driving by young males with a pattern of self-destructiveness not unconnected with suicidal behaviour. In Durkheim's classic effort to demonstrate a social phenomenon's relationship to other social phenomena—suicide to religion and economics—there was a disregard of any endogenous factors within the self-destructive organism itself. This neglect has persisted; in his review of 378 papers on the subject, Devries (1968) failed to find a single study of constitutional and physiological determinants of suicide. Struve (quoted in Devries 1968) found a comparable situation in a survey of some 500 publications.

However, this appears to be changing. Recently Struve, Klein, and Saraf (1972) reported results of a study of possible relationships between abnormal 'paroxysmal EEG and suicide ideation, suicide ideation plus attempts, and assaultive–destructive behaviour without a suicidal component'. For both males and females, there was found a positive significant association between the syndrome of suicidal behaviour and a particular pattern of brain performance.

> In just what manner paroxysmal EEG dysrhythmias influence the development and expression of suicidal impulses is not clear. It is doubtful that such dysrhythmias directly cause suicidal behaviour. One possibility is that such dysrhythmias are associated with vulnerability to the impairment of controls under stress leading to a variety of impulsive, short-sighted plans (ideation) and actions such as assaultive–destructive or suicidal behaviour. Another possibility is that the EEG patterns reflect a pathophysiology that predisposes to the development of depressive effect, of which suicide is a secondary expression. Since elevated 17-hydroxy-corticosteroid levels have been implicated as antecedents of suicidal behaviour, it is stimulating to note that high levels of adrenal steroids may be epileptogenic in some individuals and therefore associated with paroxysmal dysrhythmias and their behavioural consequence.

In a more psychological investigation Thomas and Greenstreet (1973) have been able to predict occurrence of suicide and four other disease states—mental illness, hypertension, coronary heart disease and tumour—among 1,130 medical students participating in a longitudinal study initiated in 1948 at Johns Hopkins University School of Medicine. While some of the determinants that enter into their prediction depend upon sociological factors such as closeness to parents and use of alcohol, other determinants such as resting diastolic pressure and cholesterol level are largely unrelated to explicit sociological stimuli.

The implication is not, of course, that the Durkheimian procedure is incorrect but that is it incomplete. In view of the vast increments of information in biological and medical sciences since Durkheim announced his rules of sociological method and applied them to the problem of suicide, it is scarcely surprising that the rules are no longer reliable guides to the descriptions and analysis of human social reality.

Paradoxically, the search for 'social facts' has been successful in at least one unexpected direction. Well known as the phenomenon will be to students of magic and witchcraft, the pervasiveness of the 'placebo effect' in securing recovery suggests the intricacy and reality of the link between social behaviour and bodily function. A frequent suggestion is that some 35 per cent of psychosomatic illnesses achieve placebo recovery, and that this is a predictable and widely occurring proportion. That misinforming patients is not essential for successful placebo effects is one possible conclusion of Park and Covi's (1965) finding that the placebo response among patients remained constant even when they were explicitly told they were being administered an inert substance rather than an active drug. In another suggestion of the effect of behaviour on body, Montagu (1972) has described what he calls 'sociogenic brain damage'—the consequence of impoverished infancies and early childhood: a child experiencing poor social relations, inadequate care, with few physical resources with which to play, will be marked by impairment of the brain function which will have effect, plainly enough, for adult competence. This has been decisively argued in terms of food intake (Kaplan 1972), and early stimulation (Scarr-Salapatek and Williams 1973).

Contrary evidence about behaviour has been reported recently by Jerome Kagan (1972), who studied rural Guatemalan children whose first years were describable as very deprived in both social and physical terms. On tests of competence taken among three-year-olds, these children are clearly inferior to children from rich environments in Cambridge, Mass. However, by age 11 Kagan finds there are no significant differences between the rich and poor groups, and concludes that the Piagetian hypothesis about the relative fixity of the program of childhood development is more likely to be correct than the environmentalist one which marks much of contemporary psychology and, let it be said, anthropology as well. A similar argument about the genetically programmed nature of language-learning has been made by Chomsky (1965); Lenneberg (1967) has argued the fineness of detail of biological control of language; Monod (1971) has noted that, based on his understanding of the genetic codes, it is possible that instructions for human language are contained in these codes.

In a study of correlations between the onset of menstruation and friendship and residence patterns among women in a college dormitory, Martha McClintock (1971) proposed that social interaction can have a strong effect on the menstrual cycle. 'There was a significant increase in synchronization (that is, a decrease in the difference between onset dates) among room mates ($P\angle0.0007$), among closest friends ($P\angle$

0·003) and among room mates and closest friends combined (P∠ 0·0003).' While McClintock does not or cannot identify the mechanism involved, what is intriguing in view of the concatenation of sociological and biological explanation is the fact that the close friendship patterns govern the physiological result rather than other variables such as nutrition, exposure to light, age, etc. (which were in turn ruled out by statistical test). However, if menstruation is regarded as social as well as physiological, then this becomes somewhat more understandable, as does McClintock's (1971) finding that females lacking contact with males cycled irregularly and less often than females exposed to males frequently. This recalls the Whitten effect in mice, in which females with suppressed oestrus can be made to cycle by introducing the smell of a male rat to them (Whitten 1956). Also recalled is the Bruce effect in which a pregnant mouse is induced to abort spontaneously when presented with a male mouse other than the progenitor of the foetus she is carrying (Bruce 1960).

An interesting associated finding here is of the effect of the contraceptive pill on various processes. For example, McClintock (1971) notes that menstrual synchrony simply does not occur among females who are taking birth-control pills; the pills render them immune to the social or other influences which determine the 'social fact' of synchrony. As well, Diamond, Diamond, and Mast (1972) discovered that the relationship between menstrual cycle and visual acuity did not hold for women taking nonsequential oral contraceptives. In essence, the birth-control pill wholly or in part induces social and perceptual patterns not discernible in undrugged females. This raises some intriguing questions which are also relevant to public policy. Does the pill affect women's and men's sexuality? It is known that among the other mammals the state of pregnancy implies reduced or absent sexual behaviour and this is presumably correlated with changes in endocrine-linked signals and other internally affected enthusiasms. In so far as the contraceptive pill provides the human body with the chemical information that it is already pregnant, and thus prevents pregnancy from occurring, does this have the effect of replicating in any degree the non-sexuality of the pregnant state? Clearly, among humans, a host of factors other than the endocrine affect sexual activity, yet there is a remarkable lack—given the importance of the subject and the number of drug-users involved—of information about the consequences, both personal and communal, of a regimen of functional pregnancy for relatively large numbers of presumably sexually available females. Among rhesus monkeys, Michael's (1970) finding of significantly reduced sexuality of females on the pill is accompanied by a finding of very reduced sexual interest of males in the females on the pill. Now, there are serious problems with this evidence, not the least of them being that it reflects experimental and not natural situations, and that the dosages given the animals may bear a different relationship to their systems than the human dosage

128

does for humans. Nonetheless, it remains realistic to ask whether similar effects on humans occur to those in the rhesus monkey, and to ask also what the long-term effect on the breeding system will be when large numbers of females are maintained in states of functional pregnancy even though socio-economically and psychologically they are by definition not pregnant—nor presumably willing to be so. Of relevance here are the findings about the relationship of pill-using and depression; if depression is taken as a facet of general social participation, then the problem of the interaction of somatic and social factors among contraceptive-pill-users becomes the more pressing to examine. In any event, what is pertinent to the programme of this paper is that, just as students of social behaviour have, it is argued, unduly ignored the infra-social aspects or components of behaviour, so also have workers in the pharmacological and other medical specialities too often ignored the implications for social behaviour of the chemical and other interventions they perform.

CONCLUSION

My effort here has been to review the potential mutual contribution of the natural and social sciences, and to sketch some phenomena, typically the scientific domain, from one half of this formal duality that are in fact comprehensible only in terms of method and data deriving from both. One immediate implication for social anthropologists is that, should this line of argument be a good one, then what is commonly called physical anthropology will become more fully integrated into the general structure of the discipline, and that its own concerns will thus necessarily be attuned more skilfully and keenly to the kinds of questions and answers that currently occupy social anthropologists. But this is a relatively trivial if not none the less practical step alongside the more adventurous and difficult one of assuming a unity between the sciences of the social and the sciences of the natural.

References

ARGYLE, M. 1972. Non-verbal communication in human social interaction. In: R. A. Hinde (ed.), *Non-Verbal Communication*. Cambridge: Cambridge University Press.

BARDWICK, J. M. 1971. *Psychology of Women: A Study of Bio-Cultural Conflict*. New York: Harper & Row.

BERNSTEIN, I. S. 1972. Daily activity cycles and weather influences on a pigtail monkey group. *Folia Primat*. **18**.

BLACK, F. W. 1973. Season of birth and intelligence in a sample of learning-disabled children. *J. Genetic Psychology* **123**.

BOLTON, R. 1973. Aggression and hypoglycemia among the Quolla: a study in psycho-biological anthropology. *Ethnology* **12** (3).

BRUCE, H. W. 1960. Block to pregnancy in the mouse caused by the proximity of strange males. *J. Reproduction and the Family* **1**.

BUFFERY, A. W. H., & GRAY, J. A. 1972. Sex differences in the development of spatial and linguistic skills. In: C. Ounsted and D. C. Taylor (eds.), *Gender Differences: Their Ontogeny and Significance*. Edinburgh & London: Churchill Livingstone.

CAMPBELL, B. (ed.) 1972. *Sexual Selection and the Descent of Man*. Chicago: Aldine.

CHAPPLE, E. 1970. *Culture and Biological Man*. New York: Holt, Rinehart & Winston.

CHOMSKY, N. 1965. *Aspects of the Theory of Syntax*. Cambridge, Mass.: MIT Press.

CROOK, J. H. 1973. Darwinism & the sexual politics of primates. *Atti Del Colloquio Internazionale Sul Tema: L' Origine Dell' Uomo*. Accademia Nazionale Dei Lincei, Roma, Quaderno N. 182.

DAVIS, D. E. 1970. Physiologic factors in aggressive behavior. In: C. H. Southwick (ed.), *Animal Behavior*. New York: Van Nostrand Reinhold.

DAWSON, J. L. M. 1972. Effects of sex hormones on cognitive style in rats and men. *Behavior Genetics* **2** (1).

DEARDEN, J. 1973. Sex-linked differences of political behavior: an investigation of the possibly innate origins. *Social Science Information* (UNESCO) 1975 (in press).

DEVRIES, A. G. 1968. Model for the prediction of suicidal behavior. *Psychological Reports* **22**.

DIAMOND, M., DIAMOND, A. L., & MAST, M. 1972. Visual sensitivity and sexual arousal levels during the menstrual cycle. *J. Nervous & Mental Disease* **155** (3).

EIBL-EIBESFELDT, I. 1970. *Ethology*. New York: Holt, Rinehart & Winston.

EPSTEIN, C. 1970. *Woman's Place*. Berkeley: University of California Press.

FAIRWEATHER, H., & HUTT, S. J. 1972. Gender differences in a perceptual motor skill in children. In: C. Ounsted and D. C. Taylor (eds.), *Gender Differences: Their Ontogeny and Significance*. Edinburgh & London: Churchill & Livingstone.

FIGES, E. 1970. *Patriarchal Attitudes*. London: Faber and Faber.

FOULKS, E. F. 1972. The Arctic hysterias of the North Alaskan Eskimo. Amer. Anthropology Assoc. *Anthropological Studies*, No. 10.

GARAI, J., & SHEINFELD, A. 1968. Sex differences in mental and behavioral traits. *Genetic Psychology Monographs* **77**, 2nd half, May.

GHISELIN, M. T. 1972. Darwin & evolutionary psychology. *Science* **179**.

—— 1969. *The Triumph of the Darwinian Method*. Berkeley: University of California Press.

GLUCKMAN, M. 1972. A bandwagon load of monkeys. *NY Review of Books*, 16 Nov.

GORDON, T. P., & BERNSTEIN, I. S. 1973. Seasonal variation in sexual behavior of all-male Rhesus troops. *Amer. J. Physical Anthropology* **38** (2).

GOTTIER, R. F. 1972. Factors affecting agonistic behavior in several subhuman species. *Genetic Psychology Monographs* **86** (2).

GRAY, J. 1971. Sex differences in emotional behaviour in mammals including man: endocrine bases. *Acta Psychologica* **35**.

—— 1972. Effects of septal driving of the hippocampal theta rhythm on resistance to extinction. *Psychology and Behavior* **8**.

HAMBURG, D. 1971. Recent research on hormonal factors relevant to human aggressiveness. *International Social Science J.* **23** (1).

—— 1972. Comments to second UNESCO experts' conference on human aggressiveness. Brussels, mimeo., 1972.

HOLTER, H. 1970. *Sex Roles and Social Structure*. Oslo: Universitetsforlaget.

HUTT, C. 1972a. *Males and Females*. Harmondsworth: Penguin.

—— 1972b. Neuroendocrinological, behavioural, and intellectual aspects of sexual differentiation in human development. In: C. Ounsted and D. C. Taylor (eds.), *Gender Differences: Their Ontogeny and Significance*. Edinburgh & London: Churchill Livingstone.

ISMAIL, A. A. A., DAVIDSON, D. W., LORAINE, J. A. 1972. Relationship between plasma cortisol and human sexual activity. *Nature* **237** (5353).

JOSLYN, W. D. 1973. Androgen-induced social dominance in infant female Rhesus monkeys. *J. Child Psychol. Psychiatry* **14**.

KAGAN, J. 1972. Cross-cultural perspectives on early development. Invited Address, Amer. Assoc. for the Advancement of Science, Dec. MS.

KANEBAR, S., & MUKERJEE, S. 1972. Intelligence, extraversion, and neuroticism in in relation to the season of birth. *J. Social Psychology* **86**.

KAPLAN, B. 1972. Malnutrition and mental deficiency. *Psychological Bulletin* **78** (5).

KEDENBURG, D., KEDENBURG, N., & KLING, A. 1974. Plasma testosterone and aggressive behavior in a patient population. *Archives of General Psychiatry*, in press.

KIMURA, D. 1973. The asymmetry of the human brain. *Scientific American*, March, **228** (3).

KOVACH, J. K. 1973. Soviet ethology: the first all-union conference on ecological and evolutionary aspects of animal behavior. *Behaviour* **44** (3–4).

KREUZ, L., & ROSE, R. 1972. Assessment of aggressive behavior and plasma testosterone in a young criminal population. *Psychosomatic Medicine* **34** (4).

KREUZ, L., ROSE, R., & JENNINGS, R. 1972. Suppression of plasma testosterone levels and psychological stress. *Archives of General Psychiatry* **26**.

LENNEBERG, E. H. 1967. *Biological Foundations of Language*. New York: John Wiley and Sons.

LESTER, E. P., DUDEK, S. D., & MUIR, R. C. 1972. Sex differences in the performance of school children. *Canad. Psychiatric Assoc. J.* **17**.

LEVINE, S. 1966. Sex differences in the brain. *Scientific American*, April.

LUCE, G. 1970. Biological rhythms in psychiatry and medicine. Public Health Service Publication #2088, N.I.M.H.: Bethesda, Maryland.

MCCLINTOCK, M. K. 1971. Menstrual synchrony and suppression. *Nature* **229**: 244–245.

MCGUIRE, M. T. *et al.* 1974. *The St Kitts Vervet*. Basel: S. Karger and Company.

MICHAEL, R. P., & ZUMPE, D. 1970. Rhythmic changes in the copulatory frequency of Rhesus monkeys (*Macaca mulatta*) in relation to the menstrual cycle and a comparison with the human cycle. *J. Reproduction and Fertility* **21**.

MILLETT, K. 1970. *Sexual Politics*. Garden City, New York: Doubleday; London: Hart-Davis, 1971.

MONEY, J., & EHRHARDT, A. A. 1972. *Man & Woman, Boy & Girl*. Baltimore and London: Johns Hopkins University Press.

MONOD, J. 1971. *Chance and Necessity*. New York: Knopf; London: Collins, 1972.

MONTAGU, M. F. ASHLEY 1972. Sociogenic brain damage. *Amer. Anthropologist* **74**, Oct.

MORIN, E. 1973. *Le Paradigme Perdu: La Nature Humaine*. Paris: Éditions du Seuil.

MYKYTOWYCZ, R. 1972. The behavioural role of the mammalian skin glands. *Naturwissenschaften* **59**.

PARK, L. C., & COVI, L. 1965. Nonblind placebo trial—an exploration of neurotic patients' response to placebos when their inert content is disclosed. *Archives of General Psychiatry* **12**, April.

PERSKY, H., *et. al.* 1971. Relation of psychologic measures of aggression and hostility to testosterone production in man. *Psychosomatic Medicine* **33** (3).

ROSE, R., & HOLADAY, J. 1971. Plasma testosterone, dominance rank, and aggressive behavior in male Rhesus monkeys. *Nature* **231**.

ROTHBALLER, A. B. 1967. Aggression, defense and neurohumors. In: C. D. Clemente and D. B. Lindsley (eds.), *Aggression and Defense*. Brain Function Conference, Vol. 5, Berkeley, Univ. of California.

SCARR-SALAPATEK, S., & WILLIAMS, M. L. 1973. The effect of early stimulation on low-birth-weight infants. *Child Development* **44**.

131

SELIGMAN, M. E. P., & HAGER, J. L. 1972. *Biological Boundaries of Learning*. New York: Appleton-Century-Crofts.

STRUVE, F. S., KLEIN, D. F., & SARAF, K. S. 1972. Electroencephalographic correlates of suicide ideation and attempts. *Archives of General Psychiatry* **27**, Sept.

TAUB, E., PERRELLA, P., & BARRO, G. 1973. Behavioral development after foreblim deafferentation on day of birth in monkeys with and without blinding. *Science* **181**.

THOMAS, C., & GREENSTREET, R. 1973. Psychobiological characteristics in youth as predictors of five disease states: suicide, mental illness, hypertension, coronary heart disease and tumor. *The Johns Hopkins Medical J.* **132**, Jan.

TIGER, L. 1969. *Men in Groups*. London: Nelson.

TINBERGEN, N. 1963. On the aims and methods of ethology. *Zeitschrift fur Tierpsychologie* **20**.

—— 1972. Functional ethology and the human sciences. (The Croonian Lecture.) *Proc. R. Soc. London B.* **182**.

VAN DEN BERGHE, P. L. 1973. *Age and Sex in Human Societies: A Biosocial Perspective*. Belmont, California: Wadsworth.

VESSEY, S. 1973. Night observations of free-ranging Rhesus monkeys. *Amer. J. of Physical Anthropology* **38**. March.

WASHBURN, S. L., & CIOCHON, R. L. 1973. Canine teeth: notes on controversies in the study of human evolution (mimeo.).

WASHBURN, S. L., & MCCOWN, E. R. 1973. Aggression and the evolution of the skull. Paper presented at the Amer. Anthropological Assoc.'s 72nd Annual Meeting, Nov., New Orleans, La.

WHITE, N. G., & PARSON, P. A. 1973. Genetic and socio-cultural differentiation in the aborigines of Arnhem Land, Australia. *Amer J. of Physical Anthropology* **38** (1).

WHITTEN, W. K. 1956. Modification of the oestrous cycle of the mouse by external stimuli associated with the male. *J. Endocrin.* **13**.

WILMORE, J. H. 1973. Strength, endurance and body composition of the female athlete. Paper presented at the Amer. Medical Assoc.'s 15th National Conference on the Medical Aspects of Sports, 1 Dec., Anaheim, California.

ZYLMAN, R. 1972. *Journal of Traffic Education* **20**, Oct.

Acknowledgements

I am grateful to Messrs Ian Bowers and Terence O'Keefe for their research assistance, and to the Harry Frank Guggenheim Foundation for its research support. However, of course, I am fully responsible for the views expressed in this article.

W. D. Hamilton

Innate Social Aptitudes of Man: an Approach from Evolutionary Genetics

It has become clear that, although learning has great importance in the normal development of nearly all phases of primate behaviour, it is not a generalized ability; animals are able to learn some things with great ease and other things only with the greatest difficulty. Learning is part of the adaptive pattern of a species and can be understood only when it is seen as a process of acquiring skills and attitudes that are of evolutionary significance to a species when living in the environment to which it is adapted.

Washburn, Jay, and Lancaster (1965)

SURVIVAL OF THE FITTEST

The phrase directs attention to differential survival. Darwin accepted it from Herbert Spencer as adequately expressing the idea of natural selection. While accusations of tautology seem hardly fair on this small phrase itself, it must be admitted that some descendent ideas in the theory of natural selection are open to attack. For example, the idea of measuring ability to survive and reproduce—biological 'fitness'—has undoubtedly been useful, but a slight haziness still lingers, a lack of precise and general definition: we do not know exactly what qualities natural selection is after. I think it is doubt like this rather than doubt about the reality and effectiveness of natural selection that inspires a present spirit of caution in evolutionary biology, including caution and distrust towards Spencer's ideogram whenever it renews aspirations to become a slogan.

A part of the difficulty, and the part I am mainly concerned with now, is that of saying exactly what are the things that natural selection is supposed to select. The fittest what? Is it a trait, an individual, a set of individuals bearing a trait, or bearing its determinants expressed or latent? Can it be a population, a whole species, perhaps even an ecosystem? In such a confusion of possibilities (and of fervent opinions either way) the individual organism stands out as one clear and obvious choice, with the number of its offspring as the measure of its fitness. But, beyond the problem of when to count and how to weight offspring for their ages, there is the problem that in sexual species the individual is really a physical composite of contributions from two parents and it may be composite in slightly different ways for different parts. Moreover,

Mendel's principles concerning the fair distribution of genes to gametes and fair competition of these in fertilization do not always hold, so that the set of offspring of a given individual may carry a biased sample from the composite. Does this matter? For safe conclusions, do we have to descend to the level of the individual gene, perhaps ultimately to that of changed or added parts of the replicating molecule? Or can we, on the contrary, confidently follow the consensus of biologists to a higher level, in believing that the generally significant selection is at the level of competing groups and species? I shall argue that lower levels of selection are inherently more powerful than higher levels, and that careful thought and factual checks are always needed before lower levels are neglected. In this I follow a recent critical trend in evolutionary thought (Williams 1966; Lewontin 1970; Hamilton 1972, and references in these works). Incidentally, to a biologist, a rather similar critique seems to be invited by the supposition that cultural evolution is independent of evolution in its biological substratum: to come to our notice cultures too have to survive and will hardly do so when by their nature they undermine the viability of their bearers.[1] Thus we would expect the genetic system to have various inbuilt safeguards and to provide not a blank sheet for individual cultural development but a sheet at least lightly scrawled with certain tentative outlines. The problem facing a humane civilization may be how to complete a sketch suggesting some massive and brutal edifice—say the outlines of an Aztec pyramid —so that it reappears as a Parthenon or a Taj Mahal. These ideas concerning cultural evolution will not be expanded in what follows, but I hope to produce evidence that some things which are often treated as purely cultural in man—say racial discrimination—have deep roots in our animal past and thus are quite likely to rest on direct genetic foundations. To be more specific, it is suggested that the ease and accuracy with which an idea like xenophobia strikes the next replica of itself on the template of human memory may depend on the preparation made for it there by selection—selection acting, ultimately, at the level of replicating molecules.

Returning to the problem of units of selection, Darwin himself, vague about the process of heredity, based most of his arguments on considerations of the fitness of individuals. He made occasional exceptions, as for the social insects where he treated the 'family group' as the unit of selection. I believe even these limited concessions were incautious (Hamilton 1972), and value his judgement more where, discussing the evolution of courage and self-sacrifice in man, he left a difficulty apparent and unsolved. He saw that such traits would naturally be counter-selected *within* a social group whereas in competition *between* groups the groups with the most of such qualities would be the ones best fitted to survive and increase. This open problem which Darwin left is really the starting-point of my own argument, but it is historically interesting to note that after some initial wavering between the calls of

Spencer, Kropotkin, and others, almost the whole field of biology stampeded in the direction where Darwin had gone circumspectly or not at all.

Until the advent of Mendelism uncritical acceptance of group selection could be understood partly on grounds of vagueness about the hereditary process. For example, courage and self-sacrifice could spread by cultural contagion and, in so spreading, modify heredity as well. But in the event neither the rediscovery of Mendel's work nor the fairly brisk incorporation of Mendelism into evolutionary theory had much effect. From about 1920 to about 1960 a curious situation developed where the models of 'Neo-Darwinism' were all concerned with selection at levels no higher than that of competing individuals, whereas the biological literature as a whole increasingly proclaimed faith in Neo-Darwinism, and at the same time stated almost all its interpretations of adaptation in terms of 'benefit to the species'. The leading theorists did occasionally point out the weakness of this position but on the whole concerned themselves with it surprisingly little (references in Hamilton 1964, 1971, 1972).

With facts mostly neutral and theory silent it seems that we must look to the events and the 'isms' of recent human history to understand how such a situation arose. Marxism, trade unionism, fears of 'social darwinism', and vicissitudes of thought during two world wars seem likely influences. Confronted with common social exhortations, natural selection is easily accused of divisive and reactionary implications unless 'fittest' means the fittest species (man) and 'struggle' means struggle against nature (anything but man). 'Benefit-of-the-species' arguments, so freely used during the period in question, are seen in this light as euphemisms for natural selection. They provide for the reader (and evidently often for the writer as well) an escape from inner conflict, exacting nothing emotionally beyond what most of us learn to accept in childhood, that most forms of life exploit and prey on one another.

LEVELS OF SELECTION

Often the problem is not acute. There are many traits like resistance to disease, good eyesight, dexterity which are clearly beneficial to individual, group, and species. But with most traits that can be called social in a general sense there is some question. For example, as language becomes more sophisticated there is also more opportunity to pervert its use for selfish ends: fluency is an aid to persuasive lying as well as to conveying complex truths that are socially useful. Consider also the selective value of having a conscience. The more consciences are lacking in a group as a whole, the more energy the group will need to divert to enforcing otherwise tacit rules or else face dissolution. Thus considering one step (individual vs. group) in a hierarchical population structure, having a conscience is an 'altruistic' character. But for the next step—group vs. supergroup—it might be selfish, in the sense that the groups with high levels of conscience and orderly behaviour may grow too fast and threaten to overexploit the resources on which the whole supergroup depends. As a more biological instance similar considerations apply to sex ratio, and here a considerable amount of data has accumulated for arthropods (Hamilton 1967).

A recent reformulation of natural selection can be adapted to show how two successive levels of the subdivision of a population contribute separately to the overall natural selection (Price 1972). The approach is not limited to Mendelian inheritance but its usefulness in other directions (e.g. cultural evolution) has not yet been explored.

Consider a population consisting of a mixture of particles, and suppose we are interested in the frequency of a certain kind of particle G. Suppose the particles are grouped: let the subscript s denote the sth subpopulation. For subpopulation and for the whole we define parameters relevant to natural selection as follows:

	Subpopulation	Whole population
Number of particles:	n_s	$N = \Sigma n_s$
Frequency of G:	q_s	$q = \Sigma n_s q_s / N$
Mean fitness:	w_s	$w = \Sigma n_s w_s / N$

Fitness measures the amount of successful replication of particles in one 'generation'. Thus the total population of the next generation will be $N' = \Sigma n_s w_s$. The symbol ' denoting 'next generation' is used again in the same sense in the following further addition to notation:

	Subpopulation	Whole population
Change in frequency of G in one generation:	$\Delta q_s = q'_s - q_s$	$\Delta q = q' - q$

With such notation it is easy to derive:

$$w \, \Delta q = \Sigma n_s w_s (q_s - q)/N + \Sigma n_s w_s \, \Delta q_s / N$$
$$= Covariance \, (w_s, q_s) + Expectation \, (w_s \, \Delta q_s) \tag{1}$$

where *Covariance* and *Expectation* are understood to involve weighting by the n_s as indicated. This is Price's form.[2] The covariance term represents the contribution of *intergroup* selection, so quantifying the intuitive notion that high q_s must cause high w_s for selective change to occur. The expectation term represents the contribution of *intragroup selection*. It is possible to apply the formula within itself, to expand $w_s \, \Delta q_s$. For example, if the next level is that of diploid individuals and si indexes the ith individual of the sth group we have $n_s = \Sigma 2$, $q_s = \Sigma 2 q_{si}/n_s$ and $w_s = \Sigma 2 w_{si}/n_s$ where these summations are understood to cover all i instead of all s as previously. Then $w_s \, \Delta q_s$ decomposes into two terms, one of which represents ordinary diploid selection with strictly Mendelian inheritance, while the second represents the effects of genetic 'drift' (random sampling effects), and 'drive' (non-Mendelian ratios). Even this latter term can be reformulated using (1), but then our 'groups' are the fundamental particles themselves which, neglecting mutation, must give $\Delta q_{particle} = 0$, so that here finally the second term goes out.

An often useful rearrangement of (1), which shows the dependence of selection on the variability in its units, introduces the regression co-efficient of w_s on q_s. If β_1 is this coefficient:

$$w\Delta q = \beta_1 \ Variance \ (q_s) + Expectation \ (w_s \ \Delta q_s) \qquad (2)$$

Conceptual simplicity, recursiveness, and formal separation of levels of selection are attractive features of these equations. But, of course being able to point to a relevant and generally non-zero part of selective change is far from showing that group selection can override individual selection when the two are in conflict. Moreover, even the possibility of devising model circumstances in which a positive group-selection term (first term) outweighs a negative individual selection one (second term, assuming no further levels), gives no guarantee that 'altruism' can evolve by group selection: we have to consider whether the population can get into the specified state, and, if it can, whether its present trend will continue. For example, if we suppose persistent groups with no extinction and no intergroup migration it is easy to arrange that the group-beneficial effect (β_1), of frequent altruism in a certain group is so large that the rapid expansion of the group with the highest frequency of G (q_m say) draws the population q rapidly upwards. But q will never reach or pass q_m, and must eventually fall, remaining below the ever-falling value of q_m. Admittedly, all this is reasonably obvious without the equation; but the equation does emphasize that natural selection depends on a certain variance which in this model must at last die away as the best group increasingly predominates. This is the essential objection to an algebraic model of Haldane (1932) for selection of altruism, which other writers have wrongly treated as the first successful analytical model for altruism. In verbal discussion Haldane himself admitted the necessity of a device to maintain diversity. He suggested that if groups split on reaching a certain size, random assortment of altruists and egotists would raise the frequency of altruists in some daughter moieties, and if the critical size was low enough and the group advantage of altruism high enough, a process having endless overall enrichment in altruism might be devised. Increasing the intergroup variance by random (or, better, associative) division of existing tribes leaves less variance within groups, which, as a development of the equation will shortly make plain, weakens the power of individual selection, and this further improves the case. But Price's equation does not seem to lend itself to a detailed analysis of Haldane's suggestion—indeed the lack of analysis by Haldane himself suggests that it is not easy. The value of the covariance approach lies not so much in analytical penetration as in clarifying the approach to a problem.

137

Therefore, noting hopeful auguries in Haldane's tribe-splitting no-migration idea, let us now turn to a model at the opposite extreme in which groups break up completely and re-form in each generation. Suppose that on reaching maturity the young animals take off to form a migrant pool, from which groups of n are randomly selected to be the group of the next generation. Assume completely asexual reproduction (or perfect matriclinal or patriclinal inheritance of a cultural trait in an ordinary population), and assume that an altruist gives up k units of his own fitness in order to add K units to the joint fitness of his $(n-1)$ companions. These companions are a random selection from the gene pool and therefore, in a supposed infinite population, have the expected gene frequency of the gene pool. Thus compared to a non-altruist, the altruist is putting into the next gene pool fewer of his own genes plus a random handful from the pool of the last generation. Obviously his trait is not enriching the population with genes that cause the trait. The specification of grouping has been a mere gesture. Nevertheless it is instructive to see how equation (2) handles the matter.

With asexuality individuals are basic particles, so, as already explained, the recursive use of (2) to expand its second term gives simply:

$$w \, \Delta q = \beta_1 \, Var(q_s) + E\{\beta_0 \, Var_s(q_{si})\}$$

All units are now of the same size, so *Var(iance)* and *E(xpectation)* can have their conventional meanings. Since β_0 does not vary with group constitution,

$$w \, \Delta q = \beta_1 \, Var(q_s) + \beta_0 \, E\{Var_s(q_{si})\} \tag{3}$$

and the expectation is what is commonly called the within-group variance.

With random grouping, the distribution of the different compositions of groups will be binomial with parameters (q, n). The variance of q_s is $\frac{1}{n}pq$. Likewise it is easily shown that $E\{Var_s(q_{si})\} = \frac{n-1}{n} \, p\dot{q}$, so that

$$w \, \Delta q = \frac{1}{n} p \, q \, \{\beta_1 + (n-1)\beta_0\} \tag{4}$$

This already shows the characteristically greater power of the lower level of selection as dictated by a ratio of variances that is bound to hold when grouping is random or nearly so.

In a group with v of its n members altruistic fitnesses are as follows:

Group mean	Selfish member	Altruistic member
$1 + \frac{v}{n}(K-k)$	$1 + v\frac{K}{n-1}$	$1 - k + (v-1)\frac{K}{n-1}$

138

Thus by inspection and by subtraction respectively

$$\beta_1 = K-k \text{ and } \beta_0 = -k - \frac{K}{n-1} \tag{5}$$

Substituting in (4) we find:
$$w \, \Delta q = -k \, p \, q \tag{6}$$

This confirms the earlier argument that altruism cannot progress in such a model. It seems at a first glance that the benefits dispensed by altruists have been entirely null in the working of the model, but they affect it through their involvement in mean fitness:

$$w = p + q(K-k)$$

This being the only involvement of K, we see that the most that altruism can achieve in the model is a slowing of the rate at which natural selection reduces its frequency—an effect which I explained earlier as altruism diluting each new gene mixture by adding, as it were, handfuls taken randomly from the previous one. Apart from this minor effect the model, like Haldane's algebraic one, is a failure, in spite of having shifted to the opposite extreme in respect of migration. It reveals a group selection component which is not zero but which is bound in an unchanging subordination to the individual selection component. However, the relation between the two variances in this case suggests how we must change the model to make altruism succeed: $Var(q_s)$ must be increased relative to $E\{Var_s(q_{si})\}$. As already mentioned, this can be done by making G assort positively with its own type in settling from the migrant cloud. Suppose it assorts to such a degree that the correlation of two separate randomly selected members of a group is F. If this correlation is achieved by having a fraction F of groups made pure for each type and then the remainder again formed randomly, then it is easily shown that the between and within group variances are respectively:

$$\frac{1}{n} p \, q \, (1-F+nF) \quad \text{and} \quad \frac{1}{n} p \, q \, (n-1)(1-F)$$

Putting these results and those of (5) into equation (3) we find as the generalization of (6):

$$w \, \Delta q = p \, q \, (FK-k) \tag{7}$$

so that the criterion for positive selection of altruism is

$$\frac{K}{k} > \frac{1}{F}$$

Now the model can be made to work. Moreover, the simple form and the independence of group size suggest that the criterion may hold beyond the limits of the rather artificial model discussed here. Careful thought confirms that this is indeed the case: the criterion is completely general for asexual models with non-overlapping generations, and also holds for sexual diploid models when the coefficient F is suitably redefined (Hamilton 1964, 1970). The easiest way to see the basis of generality is to notice that the benefits of altruism do not now fall on a random section of the population and therefore do not simply enlarg⟨ the existing gene pool; instead they fall on individuals *more likely to b altruists* than are random members of the population. Indeed, the exi⟨ tence of the positive correlation F could be interpreted as implying in this case that there is a chance F that the K units of fitness are definitely given to a fellow altruist, while with chance $(1-F)$ they are given (as they always were in the previous version) to a random member of the population.

The redefinition necessary for diploid organisms involves specifying a regression coefficient, b_{AB}, representing the regression of the genotype of recipient B on genotype of donor A. Often this is the same as the correlation coefficient of such genotypes (it always is so in the haploid case), but where they differ it is the regression coefficient that gives the prediction of gene content that we need. To get the form like (7) which applies to diploid selection other changes are obviously necessary, notably dividing pq by two to get the variance of gene frequency between pairs instead of that between individuals and other more complex changes connected with dominance and details of the assortative process. However, it is striking that a criterion like $(FK-k)>0$ can be shown to determine positive selection of each genotype, and can be generalized to cover cases where A distributes various effects, positive or negative, to numerous individuals B, C, D, . . . all having different regressions on A. Including A himself in the list of recipients we arrive at the idea of A's 'inclusive fitness'; his basic non-social fitness, plus all the effects caused by his action when each has been devalued by a regression coefficient.

The usefulness of the 'inclusive fitness' approach to social behaviour (i.e. an approach using criteria like $(b_{AB}K-k)>0$) is that it is more general than the 'group selection', 'kin selection', or 'reciprocal altruism' approaches and so provides an overview even where regression coefficients and fitness effects are not easy to estimate or specify. As against 'group selection' it provides a useful conceptual tool where no grouping is apparent—for example, it can deal with an ungrouped viscous population where, owing to restricted migration, an individual's normal neighbours and interactants tend to be his genetical kindred.

Because of the way it was first explained, the approach using inclusive fitness has often been identified with 'kin selection' and presented strictly as an alternative to 'group selection' as a way of establishing altruistic social behaviour by natural selection (e.g. Maynard Smith 1964; Lewontin 1970). But the foregoing discussion shows that kinship should be considered just one way of getting positive regression of genotype in the recipient, and that it is this positive regression that is vitally necessary for altruism. Thus the inclusive-fitness concept is more general than 'kin selection'. Haldane's suggestion about tribe-splitting can be seen in one light as a way of increasing intergroup variance and in another as a way of getting positive regression in the population as a whole by having the groups which happen to have most altruists divide most frequently. In this case the altruists are helping true relatives. But in the assortative-settling model it obviously makes no difference if altruists settle with altruists because they are related (perhaps never having parted from them) or because they recognize fellow altruists as such, or settle together because of some pleiotropic effect of the gene on habitat preference. If we insist that group selection is different from kin selection the term should be restricted to situations of assortation definitely not involving kin. But it seems on the whole preferable to retain a more flexible use of terms; to use group selection where groups are clearly in evidence and to qualify with mention of 'kin' (as in the 'kin-group' selection referred to by Brown 1973), 'relatedness' or 'low migration' (which is often the cause of relatedness in groups), or else 'assortation', as appropriate. The term 'kin selection' appeals most where pedigrees tend to be unbounded and interwoven, as is so often the case with man.

Although correlation between interactants is necessary if altruism is to receive positive selection, it may well be that trying to find regression coefficients is not the best analytical approach to a particular model. Indeed, the problem of formulating them *exactly* for sexual models proves difficult (Hamilton 1964). One recent model that makes more frequent group extinction the penalty for selfishness (or lack of altruism) has achieved rigorous and striking conclusions without reference to regression or relatedness (Eshel 1972). But reassuringly the conclusions of both this and another similar model (more general but less thorough and much less well explained; Levins 1970) are of the general kind that consideration of regression leads us to expect. The regression is due to relatedness in these cases, but classified by approach these were the first working models of group selection.

TRIBAL FACIES OF SOCIAL BEHAVIOUR

One of the conclusions of the models just mentioned is that with a grouped population the migration between groups is crucially important in determining the general level which altruism can reach within a group. This is something which should now seem fairly obvious but which has been surprisingly overlooked in most discussions of group selection previous to Eshel's. The less migration there is the more relatedness will build up within groups. This will permit selection of acts with low gain ratio (i.e. ratios like K/k) but the gain ratios must always exceed one, and this means that the act must actually aid group fitness in some way—reduce its chance of sudden extinction (Levins 1970;

Eshel 1972), or increase its rate of emission of migrants (Wright, mention in various papers from 1945 to 1965). With the last eventuality it is better for altruism if the migrants get together in small groups to found new colonies than if they all enter existing groups, since entering undermines the assumption of low migration—in other words reduces intergroup variance. If groups of founder migrants are assortative so much the better, although if they are so by coming all from the same parent group this could be treated as fission. Likewise if migrant acceptance is the established mode, so much the better if groups selectively accept altruists. The ability of animals to exercise such discrimination may seem dubious when behaviour even in man is rather indefinite in this respect, but it is noticeable that with many of the tight-knit groups of social carnivores and primates the would-be immigrant does go through a probationary period of hostile treatment and low status, which sometimes terminates his attempt to join (Altmann & Altmann 1972). Similar phenomena of possibly similar significance are certainly not lacking in man, witness the harsh requirements of achievement and service for an aspiring Amerindian brave (or neophyte British doctor for that matter) and the general suspicion, hostility, and low position accorded to wealthless immigrants. I should add here that the idea that such behaviour is natural in man does not mean that it is right or even sensible under modern conditions. For example, the immigrants may bring new skills and aptitudes, a point to which I return later. And as regards 'altruism', recent tribal immigrants are likely to be net importers of this precious stuff—themselves the losers when they expose their natural communistic generosity to civilized exploitation. On the other hand, when experience of ambient guile and cupidity has taught them better, such immigrants may learn to confine this generosity again among themselves and to turn outwards a contemptuous and unsympathetic attitude which is also typically tribal; but such expected ambivalence in tribal feeling is another matter to which I must return.

I have carefully spoken of 'migration' rather than 'migration rate' so far, and in doing so intend to emphasize that it is the number of acts of successful migration that is important for mean intragroup relatedness. The size of demes may matter suprisingly little. An indigenous villager may know some of his many connections with other villagers and be aware of a plexus of relationship through the misty past. What might surprise him (as it surprised me) is that relatedness as measured here (and as manifested in physical similarity) builds up just as much eventually in a large unit, say a remote town, as it does in a village, if the same actual numbers enter and leave each generation. In other words, connections which the remote townsman does not so easily know of make up in multiplicity what they lack in close degree. Of course, a large unit usually does have more migration, and consequently less intrarelatedness, but the important thing is that it is the number of

142

migrants rather than the size of colonies that determines this. For Wright's simple island model where migrants go anywhere among infinitely many colonies the approximate formula for mean intragroup relatedness (after migration has occurred) is very simple, $b = 1/(2M+1)$ where M is the number of migrants (assumed small) per subpopulation per generation. So with one migrant exchanged every other generation we find $b = \frac{1}{2}$, the same as for siblings in a panmictic population, and we therefore expect the degree of amicability that is normally expressed between siblings. If three migrants go (and three come) every generation we get $b = \frac{1}{7}$. This is slightly more than the relatedness of outbred cousins ($b = \frac{1}{8}$), so such colonies should be slightly more intra-amicable than groups of cousins would be. If, as normally happens, migrants tend to go mainly to neighbouring populations, then emigrant and immigrant genotypes will tend to correlate and so a given level of relatedness can be maintained with more migration. However, recent achievements with the analysis of the harder stepping-stone model (Kimura and Weiss 1964; Maruyama 1972), which covers the island model as a special case, show that the difference is not very great as regards own-deme relatedness. Consider the case where the colonies are supposed spread on the plane in a square lattice. Suppose that an act of migration is either 'distant', with the migrant going to any deme among the infinitely many, as in the island model, or 'close', with the migrant going to one of the four neighbouring demes, and suppose that the odds on events of the two kinds are specified. With odds 100:1 for 'distant' to 'close' $M = \cdot 5$ leads to $b = \cdot 5$ within colonies to a good approximation: in other words, the change produced by such a small amount of local migration is negligible. If the odds are reversed to 100:1, implying local migration much more probable than distant migration, the Kimura–Weiss solutions show that the relatedness only rises to $\cdot 68$. With 10 times as much migration (i.e. about 5 exchanged per deme per generation) the corresponding relatednesses are $\cdot 09$ and $\cdot 17$, so relatedness still only doubles when migration is local rather than distant. A much greater contrast is apparent in the relatedness of individuals of neighbouring demes: when distant migration preponderates this relatedness tends to be extremely small, but when close migration preponderates members of a neighbour deme can easily have more than half the relatedness that applies to an own-deme member. And up to a point increasing migration *reduces* the contrast between own and neighbour deme, so that there are genetical as well as cultural reasons why, in man, intergroup migration and marriage should decrease intergroup hostility.

Two other points seem worth making about the stepping-stone model. One is that, in the one-dimensional version of this model which could apply to demes in a linear habitat such as a coastline or river, relatedness holds up much more strongly as local migration is increased, and relatedness to neighbour–deme member more strongly still (relative to the two-dimensional case). This means that hardly any extra hostility is expected to members of neighbouring demes. From this point of view, a sea-shore phase of hominid evolution, if it occurred, should have been particularly harmonious. The other point concerns the distribution of gene frequencies. The apparent variability of colonies is expected to change rather sharply at certain critical levels of migration. These are $M = \cdot5$ for the island model and $M = 1$ for the two-dimensional stepping-stone model with close migration predominant. This means that at about the point where the colony members are related to each other like outbred sibs it should become relatively easy for individuals to detect a fairly clear difference in appearance when comparing fellow colony members with outsiders. Actually, in the stepping-stone model the possibilities with regard to patchiness and cline-like effects are complex, but, considering simultaneously several traits which are independently inherited and at most weakly selected, the complex overlap of patterns should make possible fairly accurate separation of 'us' and 'them' at the level of colonies. We shall shortly see why natural selection might favour motivation and ability so to discriminate.

What is happening to the ordinary families embedded in these supposedly endogamous colonies? Siblings, parents, and offspring will still be the individual's closest relatives. Owing to the inbreeding, their relatednesses will be above the value of $\frac{1}{2}$ that applies under outbreeding. Thus an individual should be more altruistic than usual to his immediate kin. But other neighbours who are not immediate kin are now also closely related, and it is this reduced contrast between neighbours and close kin that will give what is probably the most striking effect: we expect less nepotistic discrimination and more genuine communism of behaviour. At the boundary of the local group, however, there is usually a sharp drop in relatedness. If migrants (or whole groups) are very mobile, leading to an 'island' rather than a 'stepping-stone' situation, this drop may be such as to promote active hostility between neighbouring groups.[3] Even though these groups have some relatedness, as practical limitations to distant migration naturally ensure, the contrast is still such that a minor benefit from taking the life of an outsider would make the act adaptive. Recent studies on hunting dogs (Kuhme 1965; Lawick-Goodall & Lawick-Goodall 1971) and hyenas (Kruuk 1972), show strangers sometimes being killed, while within-the-group relations are usually amicable and even communistic. The most serious wounding which Lawick-Goodall (1971) recorded in her study of chimpanzees occurred when two males combined to attack a male of another group. Bygott (1972) witnessed a fierce attack by a group of male chimpanzees on a stranger female. The female escaped but the males caught and ate her infant. Trespassers may sometimes be killed in wolves (Mech 1972), and in rats (Lorenz 1966). In lions (Schaller 1972), langurs (Sugiyama & Parthasanathy 1969), and probably in rats and mice there is also killing of strange young, but this is probably in a rather different category because it is done only by males whose aim seems to be to sire new offspring on the mothers they bereave.

These phenomena are reminiscent of the intercolony hostility so often observed in social insects, where again actual killing may be frequent along the frontiers. With regard to relatedness, the situation is the same except that intrarelatedness in groups is usually due to all colony members being descendants of a single queen. But polygynic ants (e.g. the common red ants of the genus *Myrmica*) may approximate the breeding structure of group hunting carnivores rather well, and this tempts one to apply the super-organism concept often used for social insects to the cooperative social mammals. Such a view would compare the killing of occasional trespassers to the occasional minor wound with death of cells which occurs in the restrained fighting between, say, two individual dogs.

The basis for thinking that group-hunting carnivores are highly related within groups is the known low rate of migration and the reluctance with which migrants are accepted. Why such reluctance? The probable reason has already been touched on: the group has a cooperative job to do, necessary for its survival. This job is the hunting and killing of prey which are too large for one individual to tackle alone. The more work is invested in a task prior to its fruition the more worthwhile a parasitic option of behaviour becomes—at least, until parasites are too numerous (Hamilton 1971). And the more cooperation is involved in any endeavour the more scope there is for the inconspicuous idler. What is to stop a hunting dog from watching the hunt from afar and trotting up, by all possible short cuts, just after the prey has been killed ? Probably this has happened and probably groups over full of the offspring of such idlers have found themselves unable to kill prey and have died out. This would give a slow selection for features of pack behaviour (either cultural or genetic) that make infiltration progressively more difficult. Simply cutting down on immigration would have the desired effect through raising relatedness but, as mentioned earlier, a selective entrance requirement, with the applicant's behaviour watched through a probationary period, would be even better. It should be mentioned here that a development which closed a group's frontiers completely would probably also fail in the long run for reasons of general adaptation: complete inbreeding abandons the obviously important advantages of sexual reproduction, whatever these are (see Bischof, above).

Roughly, as we currently see it, a cunning ape-like creature once pushed boldly out from near niches now held by baboon and chimpanzee. Whether or not (as one quite plausible view holds) it first left its less enterprising cousins to take a holiday on the seashore, eventually it reappears on the African savannah participating in the Pleistocene wild-life bonanza as a group-hunting carnivore. In spite of the now-not-so-prehensile foot which it kept all the while in the door of an

145

omnivorous diet, it seems likely that this creature would have needed the same population structure as the other group-hunting carnivores and for the same but more urgent reasons. It is difficult to see what was the first factor in the escalation in cunning of this particular primate line, but the choice seems mainly between tools and language. The great benefits that these could confer to a cooperative hunter through improved technique and organization would ensure rapid selection for their development. But they would also affect the social situation in significant ways and indirectly this might escalate their selection: (a) both would provide extra cultural clues to group identification; (b) tools (and later other valued artefacts) would give further scope for parasitical behaviour, first intragroup but later between groups as well. Tools and possessions could be appropriated instead of made. With language in rapid evolution, learning, experience, and even intelligence would become increasingly open to parasitism. Meanwhile, increasing intelligence would make possible a very plastic approach to parasitical and altruistic behaviour, which in turn would increase the complexity of the semi-serious deception and coalition games which are so characteristic of behaviour within primate groups. Real rewards in food and mating are the incentive to this activity and thus escalate the selection for skill in play. The main point is that intelligence, plus (a) and (b), plus what has been explained about the real differentiation of genetical relatedness suggest the development of an explosive situation. Close frontiers to migrants a little more, or slightly increase group mobility, and it is possible to imagine the sudden success of a policy which makes any frontier incident an occasion for an attempt at violent incursion by the more populous group with losers killed, enslaved, or driven off. Successful occupation of the captured territory would soon bring the victors into contact with still less related 'stones' of the 'stepping-stone' lattice, which they could attack adaptively with even less reason for restraint. Increasing foresight would mean that a group would not necessarily wait until large enough to *need* neighbouring territory, if attacking a weak group while it is weak helped to ensure that space could be occupied as needed. Increasing ability to abstract and generalize would enable groups to reanimate their intragroup coalition games in the more serious intergroup context. The usual and firmest coalitions would be between related groups, as is the case with coalitions of individuals (usually males) in wild turkeys (Watts and Stokes 1971), lions (Schaller 1972) and chimpanzees (Lawick-Goodall 1971). Are group fights necessarily more serious for the species if, on the analogy of the super-organism, we are allowed to equate a few deaths to a minor wound? Perhaps not, and of course it is possible that making groups more aggressive would not 'melt' the lattice structure to the extent suggested. Moreover, groups might be units in super-groups that are themselves in a 'stepping-stone' lattice. In such cases warfare—for that is the behaviour we now survey—might carry over from intragroup behaviour (or itself spontaneously develop) quite orderly restrained procedures involving little loss of life.

In developing this admittedly speculative outline of certain cultural and genetic processes in tribal evolution, I confess a bias towards discovering the patterns of coalitions, warfare, language, contempt, and so on that are documented in certain remote peoples of the present day, for example the Yanomamo (Chagnon 1968) and various New Guinea highlanders (Matthiessen 1962; Rappaport 1971). Admittedly in these cases there is agriculture; it is possible to claim that most hunter–gatherers are more peaceable. For example why not aim to derive the customs of the Kalahari Bushmen? But most hunter–gatherers are certainly less peaceable than Bushmen. The record of human violence goes back far indeed, even if the earliest attributions (Dart's cases in *Australopithecus*: Ardrey 1961) are doubtful. A trace of homology with the sporadic violence of chimpanzees seems not impossible. Probably with hominids, as with chimpanzees, actual violence towards outsiders would contrast with restrained violence, or mere threats, used within the group, while within the group too there would be much sharing and cooperation. One neanderthal skeleton of Shanidar Cave had bone damage suggesting a stab wound (Solecki 1971). Another skeleton also had bone defects but of quite different implication: one forearm was lacking, perhaps from birth, certainly for a long time, and a healed injury to the skull showed that one eye was blind. Goodall's chimpanzees were part hostile, part sympathetic, and part indifferent to comrades suddenly crippled with polio: they did nothing positive to support them. In contrast, the neanderthals of Shanidar evidently supported a cripple, and on his death they buried him in the cave where, in other graves, they also sometimes buried their dead with flowers. These hints of violence and loyalty and (perhaps most purely human) of incipient love of things for themselves evoke a startlingly familiar and sympathetic portrait. Considering only the same affectional attributes in the present-day tribal and pastoral Kurds (as opposed to attributes connected to the ever-accelerating change in material culture), a recording angel perhaps notes today much the same events in Shanidar Cave as he noted an ice age ago (Hamilton 1937; Solecki 1971).

Probable instances of cannibalism in *Sinanthropus* and *Homo neanderthalensis* have been plausibly compared to similar recent cannibalism in New Guinea (Blanc 1961). In New Guinea, it is interesting to note, this practice acted as a kind of population control, since by eating the brain of his victim, it was believed, the headhunter won a name for a child of his own—in effect, won a birthright. Other usually less drastic beliefs and practices affecting fertility that are widespread in human cultures may help to explain how they manage to be as peaceful as they are. In so far as the practices amount to effective birth control, they cut warfare at its demographic root. Unfortunately, it is possible that in doing so they also cut an important link that has escalated the selection for intelligence (Keith 1945; Bigelow 1969; Alexander 1971; Andreski 1971). No hunt needs quite so much forethought or ability to communicate complex instructions as does a war, nor do such drastic demographic consequences hinge on the outcome.

The rewards of the victors in warfare obviously increase for peoples past the neolithic revolution. There are tools, livestock, stores of food, luxury goods to be seized, and even a possibility for the victors to impose themselves for a long period as a parasitical upper class. Hunter–gatherers, on the other hand, at most win only mates and land. It might seem that these things would not repay the expected cost of the fighting, but it has to be remembered that to raise mean fitness in a group either new territory or outside mates have to be obtained somehow. The occurrence of quasi-warlike group interactions in various higher primates (Kummer 1968; Sugiyama & Parthasanathy 1969; and references in Bigelow 1969) strongly suggests that something like warfare may have become adaptive far down in the hominid stock. These primate examples suggest the prototype war party as an all-male group, brothers and kin, practised as a team in successful hunting and at last redirecting its skill towards usurping the females or territory of another group. Out of such cells can be built the somewhat less stable organism of the post-neolithic army. The Homeric *Iliad* gives a vivid inside view of the process of coalition, while the siege it describes emphasizes the existence of economic surpluses supporting the warriors on both sides (something hunter–gatherer warriors would never have). If the male war party has been adaptive for as long as is surmised here, it is hardly surprising that a similar grouping often reappears spontaneously even in circumstances where its present adaptive value is low or negative, as in modern teenage gangs (Patrick 1973).

Whether or not the neolithic revolution brought an increase in the *per capita* incidence of violence it does seem that from then on warfare looms larger in the affairs of men. The situation seems reflected in the fact that only one of various series of pre-neolithic cave paintings depicts warfare (Pericot 1961), whereas for most early civilizations the earliest known written records are records of warfare, booty, captives, and the like.

It has been argued that warfare must be a pathological development in man, continually countered by natural selection, and this claim is sometimes based on a sweeping *a priori* view that habits of mortal intraspecific fighting must always endanger the survival of a species (Lorenz 1966). While endorsing such a view as regards wars between the few frightfully armed super-powers of today, I see no likelihood for it as regards fighting of individuals or of groups up to the level of small nations. Of course, for the species as a whole, and in the short term, war is detrimental from the biological demographic point of view, but, as shown above and elsewhere, detriment to the species does not mean that a genetical proclivity will not spread. Anyway, what is bad at one level may be good at another and the cost to the species may be paid in the long run. The gross inefficiency of warfare may be just what is necessary, or at least an alternative to birth control and infanticide, in order to spare a population's less resilient resources from dangerous exploitation. Maybe if the mammoth-hunters had attacked each other more and the mammoths less they could be mammoth-hunters still. And the rich ice-age fauna of the Americas might have had time to adapt to the human predator as it adapted in Africa if fighting had induced man to draw his curtain of overkill across the continent less rapidly. Many examples in the living world show that a population can be very successful in spite of a surprising diversion of time and energy into aggressive displays, squabbling, and outright fights. The examples range from bumble bees to European nations. In case all this reads like a paean for fascism let me add one caution from the geological record. Arms and armour seem to weigh one down in the end: it is hard in the modern world to see warfare as a stabilizing influence for man.

The relatively peaceable Bushmen may tell us something valuable about the aetiology of wars, but I am doubtful if they tell us much about the role of this factor in the main stream of human evolution. However, it is noteworthy that the Bantu who replaced the khoisaniform races in much of the rest of Africa were warlike in ways that evoke comparisons from the dark ages of Europe. The Bantu were, of course, mainly pastoralists and agriculturalists, for whom, as stated above, booty would be an important additional incentive to warfare. Pastoralists tend to be particularly warlike and the histories of civilization are punctuated by their inroads. Pastoral tribes have to be mobile in following or driving their herds and this mixes tribes and reduces relatedness of neighbours. Viewed as booty, the mobility of stock is a great convenience. Both factors must contribute to the warlike propensity.

The incursions of barbaric pastoralists seem to do civilizations less harm in the long run than one might expect. Indeed, two dark ages and renaissances in Europe suggest a recurring pattern in which a renaissance follows an incursion by about 800 years. It may even be suggested that certain genes or traditions of the pastoralists revitalize the conquered people with an ingredient of progress which tends to die out in a large panmictic population for the reasons already discussed. I have in mind altruism itself, or the part of the altruism which is perhaps better described as self-sacrificial daring. By the time of the renaissance it may be that the mixing of genes and cultures (or of cultures alone if these are the only vehicles, which I doubt) has continued long enough to bring the old mercantile thoughtfulness and the infused daring into conjunction in a few individuals who then find courage for all kinds of inventive innovation against the resistance of established thought and practice. Often, however, the cost in fitness of such altruism and sub-

149

limated pugnacity to the individuals concerned is by no means meta-phorical, and the benefits to fitness, such as they are, go to a mass of individuals whose genetic correlation with the innovator must be slight indeed. Thus civilization probably slowly reduces its altruism of all kinds, including the kinds needed for cultural creativity (see also Eshel 1972).

RECIPROCATION AND SOCIAL ENFORCEMENT

The last suggestion is rather different from saying, as has sometimes been said, that civilization selects against all kinds of creative intelli-gence. It seems to me that there are some aspects of innate intelligence that civilization steadily promotes. Mercantile operations, for example, are an inseparable part of Old World civilizations and need complex models in the minds of their operators, just as military ventures do. The main difference is in more emphasis on prudence and less on daring. It is probable that civilization has given steady selection for the intelligence needed for this mercantile kind of preparatory modelling. The intelligence that gives a good appreciation of the real principles involved in a new technology, as opposed to seeing it as a kind of magic, is probably also constantly favoured, since improvers of a technology avoid the arrows of contempt and penury that face pioneers and can do very well. However, my main reason for turning to the sub-ject of trade is to introduce the idea of another kind of positive social arrangement which thrives in a mercantile and technological atmo-sphere, for which intelligence is more necessary and relatedness much less so. This refers of course to reciprocation.

Starting perhaps with something like the meat-sharing of chimpanzees ('feed me while you have plenty and I'll feed you when I have plenty'), proceeding through barter (where differing aptitudes may begin to be important), reciprocative activity branches out into all the various business-like arrangements of modern man. The key words are 'client' and 'partner' as opposed to 'kinsman' and 'friend'.

Establishing a basis for reciprocating has problems of natural selec-tion closely similar to those of altruism as discussed so far. It is very frequently necessary for one party to execute his half of a bargain without any way of being certain that the other party will later stick to his. The best response if the other does not reciprocate is to cut off any further benefits to him (Trivers 1971). Unfortunately, this leaves the selfish non-reciprocator better off than the 'altruistic' initiator and unless the two are related this is against the habit of reciprocation, at least when the trait is rare (see discussion of Prisoner's Dilemma in Hamilton 1971; also Boorman & Levitt 1973). However, this initial barrier to selection is a slight one if the rewards of the interaction are high, as they would be when an advantageous exchange can be repeated many times (Trivers 1971; Boorman & Levitt 1973). Once the barrier is passed by genetic drift or the like, non-reciprocation finds itself in the category of maladaptive spite—harming the self to harm others more. Nepotistic altruism, of course, also has an initial barrier to pass when it first occurs by mutation. But once positive selection supervenes the resemblance between the two situations fades: reciprocal altruism of the kind described is less purely altruistic. Indeed, the term altruism may be a misnomer: there is an expectation of benefit of the initiating individual, not just an expectation of benefit to the genotype. To put the matter another way, reciprocal altruism can never be suicidal, whereas suicidal nepotistic altruism can and has evolved—it is apparent, for example, in worker sacrifices in the social insects.

Whether reciprocation involves altruism or not, we see that in so far as it involves repeated acts between the same two individuals this useful and immensely variegating type of interaction can spread genetically, given only an ability to remember individual faces of those who have helped and those who have cheated in the past. Unfortunately, by the very aid it gives to the growth and diversification of social systems, reciprocation tends to undermine the basis of its success. Situations demanding reciprocation just once between individuals destined never to meet again naturally become more common and it becomes easier for cheaters to specialize in these and to hide from retribution. Cheating can also become more subtle, especially along lines which make it hard for victims to be sure just who has cheated them.

In considering this problem I think there may be reason to be glad that human life is a 'many-person game' and not just a disjoined collection of 'two-person games'. Admittedly, it may not seem so at first. At first reading the theory of many-person games may seem to stand to that of two-person games in the relation of sea-sickness to a headache. But given also a little real intragroup altruism endowed from the tribal past, it may turn out that the one is at least a partial cure for the other (see *Figure 1*). The idea here is that for pairs in isolation the problem of cheating in a single exchange may be insoluble, and that therefore we have all evolved, more or less in proportion to our exposure to civilized (i.e. relatively panmictic) conditions, into potential cheaters. But at the same time we also have every reason to agree as to the parasitical nature of cheating as it affects the welfare of the community as a whole and to deplore its successful practice by others. So detection of cheating arouses indignation in everyone except the accused, and everyone sees a benefit to both group and self in trying to punish the cheater and in forcing restitution (some part of which, as an added incentive, may be diverted as a fee for those who administer the collective justice). The reason I believe that a little real intragroup altruism is also necessary for the evolution of efficient justice on these lines is that individuals must feel the difference between the usefulness of this behaviour and the futility of using collective power arbitrarily in ways profitless to the group. A healthy society *should* feel sea-sick when

Figure 1   Justice and the 'Prisoner's Dilemma' [4]

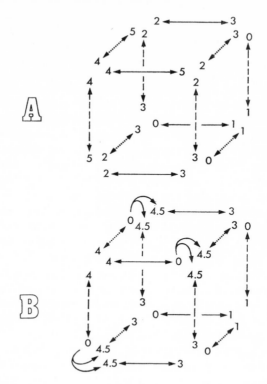

A shows a 'three-person prisoner's dilemma'. B shows the expected modification of this when justice can be enforced by a numerical majority. In every case a player has two strategies, 'cooperative' and 'selfish', and decisions between them are indicated by double-ended arrows, which are solid, dashed, or dotted to correspond to identities of the three players. Arrow ends tending towards the near upper left corner of the cube represent a decision to be cooperative; arrow ends tending to the far lower right corner of the cube represent a decision to be selfish. Numbers at the ends of arrows show the players' payoffs. Thus, naming the players in the obvious way, if 'solid' plays cooperatively but 'dash' and 'dot' both play selfishly the out-come is represented at the far lower left corner of the cube: 'solid' get zero, while 'dash' and 'dot' get three each.

In case A, payoffs are arranged so that every face of the cube presents a 'prisoner's dilemma' to each of the three possible pairs of players. Thus whatever the other pair is doing a player does best by playing selfishly; as in prisoner's dilemma the rational yet 'paradoxical' result of this is that the players all play selfishly and get one unit each, whereas had they all played cooperatively they could have four units each.

In case B, it is assumed that if two players play cooperatively while a third is selfish, the two use their united strength to 'punish' the third. His selfishly-gained five units are taken from him and divided between the two cooperators. This creates a situation where no player sees an advantage in departing from triple cooperation. Likewise no player has an advantage in departing unilaterally from the triply sefish corner; but now any two players can decide to depart from this corner by a coalition which is not only advantageous (as it was also in case A) but also secure in that neither member can benefit by defaulting.

confronted with the endless internal instabilities of the 'solutions', 'coalition sets', etc., which the theory of many-person games has had to describe. One hears that game theorists, trying to persuade people to play even two-person games like 'Prisoner's Dilemma', often encounter exasperated remarks like: 'There ought to be a law against such games!' Some of the main points of this paper can be summarized as an answer to this comment: that often, in real life, there is a law, and we can see why, and that sadly we also see the protean nature of this Dilemma, which, when suppressed at one level, gathers its strength at another.

Notes

1   Fox (1967) has also emphasized this in discussing human kinship systems and why some which are easily conceivable never actually occur. His discussion of the incest taboo is also very pertinent to the idea that follows.

2   Price (1970) first pointed out the generality and usefulness of this relation but earlier partial recognition of it seems to be due to Robertson (1966).

3   As pointed out elsewhere (Hamilton 1970, 1971), even spiteful behaviour, harming oneself in order to harm another more, is a theoretical possibility. The mean relatedness to the entire species population other than self is $-1/(N-1)$ where N is the population. If by inbreeding or otherwise a colony has grouped n identical genotypes together, then relatedness to the average outsider is $-1/(N-n)$. Thus with only a few large long-isolated groups spite is more possible.

4   The game-theoretic situation known as Prisoner's Dilemma has as paradigm the dilemma of a criminal in custody who is offered the inducement of a light sentence if he will give state evidence to clear up an important crime, while knowing that a confederate, also in custody, has the same offer and that the pay-off situation which both face is as follows, arranged from lightest to heaviest sentence:

(c) I confess, he doesn't $>$ (a) We neither confess $>$ (d) We both confess $>$
(b) He confesses, I don't.

So that (c) can be better than (a) it is assumed that they can certainly be convicted of some minor offence whereas at least one confession is needed in order to settle the major crime. For further information on this 'game' see Rapoport & Chammah (1965).

References

ALEXANDER, R. D. 1971. The search for an evolutionary philosophy of man. *Proceedings of the Royal Society of Victoria* **84**: 99–120.

ALTMANN, S. A., & ALTMANN, J. 1972. *Baboon Ecology, African Field Research.* New York: S. Karger.

ANDRESKI, S. 1971. The case for war. *Science Journal* **7** (1): 89–92.

ARDREY, R. 1961. *African Genesis.* London: Collins.

BIGELOW, R. 1969. *The Dawn Warriors.* London: Hutchinson.

BLANC, A. C. 1961. Some evidence on the ideologies of early man. In: S. L. Washburn (ed.), *The Social Life of Early Man.* London: Methuen.

BOORMAN, S. A., & LEVITT, P. R. 1973. A frequency dependent natural selection model for the evolution of social cooperation networks. *Proceedings of the National Academy of Sciences* (USA) **70**: 187–189.

BROWN, J. L. 1973. Alternative routes to sociality in jays. *American Zoologist* **12**: 642.

BYGOTT, J. D. 1972. Cannibalism among wild chimpanzees. *Nature* **238**: 410–411.

CHAGNON, N. A. 1968. *Yanomamo: The Fierce People.* New York: Holt, Rinehart & Winston.

ESHEL, I. 1972. On the neighbour effect and the evolution of altruistic traits. *Theoretical Population Biology* **3**: 258–277.

FOX, R. 1967. *Kinship and Marriage.* London: Penguin.

HALDANE, J. B. S. 1932. *The Causes of Evolution.* New York City: Cornell University Press, 1966.

HAMILTON, A. M. 1937. *Road through Kurdistan.* London: Faber and Faber.

HAMILTON, W. D. 1964. The genetical evolution of social behaviour. I and II *Journal of Theoretical Biology* **7**: 1–52.

—— 1967. Extraordinary sex ratios. *Science* **156**: 477–488.

—— 1970. Selfish and spiteful behaviour in an evolutionary model. *Nature, London* **228**: 1218–1220.

—— 1971. Selection of selfish and altruistic behaviour in some extreme models, Chapter 2, pp. 57–91. In: J. F. Eisenberg and W. S. Dillon (eds.), *Man and Beast: Comparative Social Behaviour.* Washington, D.C.: Smithsonian Press.

—— 1972. Altruism and related phenomena, mainly in social insects. *Annual Review of Ecology and Systematics* **3**: 193–232.

KEITH, A. 1946. *Essays on Human Evolution.* London: Watts.

KIMURA, M., & WEISS, G. H. 1964. The stepping stone model of population structure and the decrease of genetic correlation with distance. *Genetics* **49**: 561–576.

KUHME, W. 1965. Freilandstudien zur Soziologie des Hyänenhundes (*Lycaon pictus lupinus* Thomas 1902) *Zeitschrift für Tierpsychologie* **22**: 495–541.

KUMMER, H. 1968. *Social Organization of Hamadryas Baboons.* Chicago: University of Chicago Press.

KRUUK, H. 1972. *The Spotted Hyena.* Chicago: University of Chicago Press.

VAN LAWICK-GOODALL, H., & VAN LAWICK-GOODALL, J. 1971. *Innocent Killers.* London: Collins.

VAN LAWICK-GOODALL, J. 1971. *In the Shadow of Man.* London: Collins.

LEVINS, R. 1970. Extinction, pp. 77–107. In: M. Gerstenhaber (ed.), *Some Mathematical Questions in Biology.* Providence, Rhode Island: The American Mathematical Society.

LEWONTIN, R. C. 1970. The units of selection. *Annual Review of Ecology and Systematics* **1**: 1–18.

LORENZ, K. 1966. *On Aggression.* London: Methuen.

MARUYAMA, T. 1972. Analysis of population structure, II. Two-dimensional stepping stone models of finite length and other geographically structured populations. *Annals of Human Genetics* **35**: 182–196.

MATTHIESSEN, P. 1962. *Under the Mountain Wall.* London: Heinemann.

MAYNARD SMITH, J. 1964. Group selection and kin selection. *Nature.* London **201**: 1145–1147.

MECH, L. D. 1972. *The Wolf: the Ecology and Behaviour of an Endangered Species.* New York: The Natural History Press.

PATRICK, J. 1973. *A Glasgow Gang Observed.* London: Eyre Methuen.

PERICOT, L. 1961. The social life of Spanish palaeolithic hunters as shown by levantine art. In: Washburn, S. L. (ed.), *The Social Life of Early Man.* London: Methuen.

PRICE, G. R. 1970. Selection and covariance. *Nature, London* **227**: 520–521.

—— 1972. Extension of covariance selection mathematics. *Annals of Human Genetics* **35**: 485–490.

RAPOPORT, A., & CHAMMAH, A. M. 1965. *Prisoner's Dilemma.* Michigan: University of Michigan Press.

RAPPAPORT, R. A. 1971. The flow of energy in an agricultural society. *Scientific American* **224** (3): 116–132.

ROBERTSON, A. 1966. A mathematical model of the culling process in dairy cattle. *Animal Production* **8**: 95–108.

SCHALLER, G. 1972. *The Serengeti Lion: a Study of Predator–Prey Relations.* Chicago: University of Chicago Press.

SOLECKI, R. S. 1971. *Shanidar. The Humanity of Neanderthal Man.* London: Allen Lane.

SUGIYAMA, Y., & PARTHASANATHY, M. D. 1969. A brief account of the social life of Hanuman langurs. *Proceedings of the National Institute of Sciences, India* **35**, B: 306–319.

TRIVERS, R. L. 1971. The evolution of reciprocal altruism. *Quarterly Review of Biology* **46**: 35–57.

WASHBURN, S. L., JAY, P. C., & LANCASTER, B. 1965. Field studies of Old World monkeys and apes. *Science* **150**: 1541–1547.

WATTS, C. R., & STOKES, A. W. 1971. The social order of turkeys. *Scientific American* **224** (6): 112–118.

WILLIAMS, G. C. 1966. *Adaptation and Natural Selection, a Critique of some Current Evolutionary Thought.* Princeton, N.J.: Princeton University Press.

WRIGHT, S. 1945. Tempo and mode in evolution: a critical review. *Ecology* **26**: 415–419.

—— 1965. Factor interaction and linkage in evolution. *Proceedings of the Royal Society, B,* **162**: 88:–104.

Biographical Notes

BISCHOF, NORBERT. Born 1930, Germany; educated University of Munich, Diplom-Psychologie, Dr. Phil., Privatdozent.

Lecturer in Experimental Psychology, University of Munich, 1966–73; Research Associate, Max-Planck-Institut für Verhaltensphysiologie, Seewiesen, 1958–73; Sherman Fairchild Distinguished Scholar, California Institute of Technology, 1973– .

Author of 'Erkenntnistheoretische Grundlagenprobleme der Wahrnehmungs-psychologie', 'Psychophysik der Raumwahrnehmung', and 'Stellungs-, Spannungs-und Lagewahrnehmung' in *Handbuch der Psychologie*, Vol. 1/I *Bewusstsein und Wahrnehmung*, 1966; with E. Scheerer, 'Systemanalyse der optisch-vestibulären Interaktion bei der Wahrnehmung der Vertikalen', 1970; 'Inzestbarrieren in Sauge-tiersozietäten' (*Homo*, 1972); 'A systems approach towards the functional connec-tions of fear and attachment', in: M. Lewis and L. Rosenblum (eds.), *The Origins of Human Behavior* (in press); *Das Rätsel Œdipus* (in prep.); with H. Böttger, 'Untersuchungen zur Familienauflösung bei Wildgänsen' (*Zeitschrift für Tier-psychologie*, in prep.); with T. Schottenloher, 'Group structure and motivational structure in mammalian societies'.

BLURTON JONES, NICHOLAS. Born 1936, England; educated University of Reading, B.Sc. General (Zoology and Psychology), B.Sc. Special (Zoology); Oxford Univer-sity, D.Phil. (Animal Behaviour).

Departmental Research Officer, Department of Zoology, Oxford, 1962–63; Lecturer in Psychology, University of Bristol, 1963–64; Lecturer in Developmental Ethology, University of London, Institute of Child Health, 1964–72; Fulbright Fellow, Johns Hopkins University, 1965–66; Senior Lecturer in Developmental Ethology, Institute of Child Health, 1972– .

Editor of *Ethological Studies of Child Behaviour*, 1972; *Observational and Experi-mental Studies of Threat Displays of the Great Tit*, 1968. Author (with M. J. Konner) of 'Sex differences in behaviour of Bushman and London children' in *Social Behaviour and Ecology of Primates*, ed. Michael and Crook, 1973.

CHANCE, MICHAEL ROBIN ALEXANDER. Born England, 1915; educated University College London (London University), B.Sc. (Zoology); Birmingham University, Ph.D. (Medicine), D.Sc.; F.Inst.Biol.

Lecturer in Experimental Pharmacology, Birmingham University Medical School, 1946–66; Director of the Ethology Laboratories, Uffculme Clinic, Birming-ham University Medical School, 1958– ; Reader in Ethology, University of Birm-ingham, 1966– ; Visiting Professor in Ethology and Anthropology, Rutgers University, 1968–69.

Author of *Social Groups of Monkeys, Apes and Men* (with Clifford Jolly), 1970; 'The nature and special features of the instinctive social bond of the primates' in *Social Life of Early Man*, ed. S. L. Washburn, 1961; 'Mammalian behaviour

157

studies in medical research', *The Lancet*, 5 Oct. 1957; 'Social behaviour and primate evolution' in *Culture and the Evolution of Man*, ed. A. Montagu, 1962; 'Ethology and psychopharmacology' in *Psychopharmacology: Dimensions and Perspectives*, ed. C. R. B. Joyce, 1968; 'Attention structure as the basis of primate rank orders', *Man*, vol. 2, no. 4, 1967; and over sixty papers on pharmacology, ethology, and anthropology.

FOX, ROBIN. Born 1934, England; educated London School of Economics (London University), B.Sc. (Sociology); Harvard University; London University, Ph.D.

Tutor in Social Relations, Harvard University, 1958–59; Lecturer in Sociology with special reference to Social Anthropology, University of Exeter, 1959–63; Lecturer in Social Anthropology, London School of Economics, 1963–67; Professor of Anthropology, Rutgers University 1967–   ; National Institute of Mental Health Special Fellow, Stanford University Medical School, 1969–70; Director of Research, H. F. Guggenheim Foundation, 1972– .

Author of *Kinship and Marriage: An Anthropological Perspective*, 1967; *The Keresan Bridge: A Problem in Pueblo Ethnology*, 1967; (with Lionel Tiger) *The Imperial Animal*, 1971; *Encounter with Anthropology*, 1973; *The Tory Islanders: Studies in their Social Structure* (in press).

HAMILTON, WILLIAM DONALD. Born 1936, Cairo, Egypt; educated St John's College, Cambridge, B.A. (Natural Sciences, Genetics); London School of Economics and University College London, Ph.D.

Lecturer in Genetics, Imperial College, London University, 1964– .

Author of 'The evolution of altruistic behavior', *American Naturalist*, vol. 97, 1963; 'Extraordinary sex ratios', *Science*, vol. 156, 1967; 'Altruism and related phenomena, mainly in social insects', *Annual Review of Ecology and Systematics*, vol. 3, 1972; and other papers.

TIGER, LIONEL. Born 1937, Canada; educated McGill University, B.A., M.A. (Sociology); London School of Economics (London University), Ph.D. (Political Sociology).

Teaching Fellow, McGill University, 1957–59; Visiting Lecturer, University of Ghana, 1961; Assistant Professor of Sociology, University of British Columbia, 1963–68; Canada Council Killam Bequest for Inter-Disciplinary Research, Fellow, 1968–70; Associate Professor of Anthropology, Rutgers University, 1968–72; Professor of Anthropology, Rutgers University, 1972–   ; Director of Research, H. F. Guggenheim Foundation, 1972– .

Author of *Men in Groups*, 1969; (with Robin Fox) *The Imperial Animal*, 1971; 'Bureaucracy and urban symbol systems' in *The City in Modern Africa*, ed. Horace Miner, 1967; 'Dominance in human society', *Annual Review of Ecology and Systematics*, vol. 1, 1970; 'Biology, rhetoric, reform: the allure of low-born high ideals', *Social Science Information*, vol. 12, 1973; (with Joseph Shephee) *Women in the Kibbutz*, 1975.

Name Index

Compiled by James Urry

Subject Index

Compiled by James Urry

laughter among, 78
London nursery, 76–7
mammalian, 82–3, 85–8
play among, 76, 77
rearing of, 70, 73–4, 85–6, 87
psychology of, 72, 73, 127
socialization of *see* socialization
Zhun/twa (!Kung) Bushman, 78, 79, 88
infants, 70, 73
China *see* Taiwan
circadian rhythms, 3, 117–18
codes
Bernstein on linguistic, 100–2, 103–4
genetic, 122, 127
communication
aggression as *see* aggression
attention structures and *see* attention structures
by display *see* display
by gesture *see* gesture
human, 101–4
non-verbal, 93, 100–1, 104, 110
see also linguistics
primate, 93, 94–5, 116
theory, 93
contact, bodily as communication, 94, 96–7, 100, 109–10
see also display *and* gesture
contraception and menstruation, 128–9
culture
factors of compared to nature, 9, 115, 117, 134
incest and, 37, 38, 41, 61–2

Darwinian approaches
neo-, 1, 2, 5, 135
social, 135
demography, 1, 70, 135
dermatology, Australian aboriginal, 117
disease
hereditary, 56, 135
prediction of, 126
display
social forms associated with, 93, 94–5, 96–8, 105, 110
see also gesture *and* threats
dominance, 4, 93, 94, 95, 99, 102, 105, 123
family organization and, 20
mating and, 12, 22, 26, 51
see also submission
domestication, animals and incest after, 56

ecology
childbirth and, 118
ethology and, 70, 83, 116
evolution and, 145–6, 149
human, 79–80, 83–4, 85, 87–8
intelligence and, 101–2, 103–4
palaeontology and, 3
primate social structures and, 12, 13, 15, 24, 29–30, 83, 85, 87–8, 94–5
study of, 2
embryology, 70
neuro-, 69, 84
endocrines, 118, 121–5
sex differences and, 119
see also hormones
endocrinology, 3, 6, 121
Eskimo, North Alaskan and hysteria among, 119
ethnobiology, 118
ethology, 1–2, 70–2
anthropology and, 69–70, 71–2, 72–3, 80–1, 88–9
ecology and, 70, 83–4, 85–6, 87–8, 116
methodology of, 71–82
eugenics, 1 *see also* genetics
evolutionary theory, 1, 72, 84, 116, 121, 133–5
biology and, 2, 133, 135
cultural, 136
Darwin on, 1, 2, 5, 133–5
genetical, 3
Mendel and, 134, 135, 136, 137
human, 3, 5, 70, 85–6, 89, 145–6, 147–9 *see also* palaeontology, human
selection and, 57–9, 133, 135–9
socialization and, 76
exogamy, 12, 21, 28
selection advantages and, 57–8

family
incest and, 59–60, 61–2
nuclear, 4, 24, 28
origin of, 10, 29, 39
polygynous, 12, 14, 24
primate, 12, 15, 16, 17, 18, 21, 29–30
feeding, primate, 16, 18, 24
fossils *see* palaeontology
friendship, 115

games, 10 *see also* play
genetics, 1, 30, 117
aboriginal, 117
codes and, 122, 127
'drift' and, 136–7